Naturalists' Handbooks 12

T0198071

Animals of the surface film

MARJORIE GUTHRIE
Honorary Research Fellow in Zoology,
Manchester University

With illustrations
and plates by
Peter Hayward

Richmond Publishing Co. Ltd.

P.O. Box 963, Slough, SL2 3RS, England

Series editors
S. A. Corbet and R. H. L. Disney
Advisory board
J. W. L. Beament, V. K. Brown,
J. A. Hammond, A. E. Stubbs

Published by The Richmond Publishing Co. Ltd.,
P.O. Box 963, Slough, SL2 3RS
Telephone Farnham Common (02814) 3104

© The Richmond Publishing Co. Ltd. 1989

ISBN 0 85546 271 X Paper
ISBN 0 85546 272 8 Hardcovers

Printed in Great Britain

Contents

Editors' preface

Students at school or university, and others without a university training in biology, may have the opportunity and inclination to study local natural history but lack the knowledge to do so in a confident and productive way. The books in this series offer them the information and ideas needed to plan an investigation, and practical guidance to help them carry it out. They draw attention to regions on the frontiers of current knowledge where amateur studies have much to offer. We hope readers will derive as much satisfaction from their biological explorations as we have done.

This book is the first to attempt to deal with the whole little-known community of freshwater surface-dwelling animals. We do not yet know the limits of that community, and anyone who is stimulated by this book to explore it is likely to discover neustonic species not described here. We would welcome information that will help any future editions of the book to give more comprehensive coverage of the neuston.

The keys are an important feature of the books in this series. Even in Britain, the identification of many groups remains a barrier to ecological research because experts often write keys for other experts, and not for general ecologists. The keys in these books are meant to be easy to use. Their usefulness depends very much on the illustrations, the preparation of which was assisted by a grant from the Natural Environment Research Council.

We are very grateful to Sir James Beament, Dr Mark Rodger, Dr William Kirk and Dr Laurie Friday for advice and help.

S.A.C.
R.H.L.D.
April 1988

Acknowledgements

My best thanks are due to Dr Peter Hayward for his fine illustrations, and his cheerful co-operation in producing them; and to Dr Sally Corbet, for suggesting such a project and helping to carry it through with much patient and useful advice, and critical attention to details. Dr Henry Disney produced the keys to larval and pupal Diptera, and made a substantial contribution to the key to adult Diptera, as well as useful comments on the Diptera generally. The key to the Coleoptera is largely the work of Dr Laurie Friday. Sir James Beament, F.R.S., provided suggestions for practical investigation of contact angles and surfactant effects. Dr Colin Johnson, Keeper of Entomology at Manchester Museum, has provided many kinds of assistance, including access to the excellent collections. My husband, Simon, has given me much encouragement and help, particularly with collecting live material for the illustrator, and with the keys to Hemiptera and adult Diptera. I am most grateful to them all; also to Drs John Dalingwater, Glynn Evans, Geoffrey Fryer and Alan Kaye for valuable discussion of various relevant topics. Finally, it must be said that to attempt to produce a guide to a habitat supporting representatives of such diverse taxonomic groups must inevitably result in many omissions, for which I am entirely responsible. My hope is that it may encourage others to derive as much pleasure as I have had from trying to find out more about these fascinating creatures.

M.G.G.

1 Introduction

A calm water surface on any aquatic habitat can support a community of animals and plants of great interest. Because similar constraints apply to all surface waters, members of the same groups of organisms may be found in like habitats in many parts of the world, even though the land surrounding them may be very different. Surface-dwelling communities may be found on most bodies of water, including small tarns and ponds, lakes, the still backwaters of rivers and even the brackish tidal pools of estuaries and saltmarshes. Protection from wind and waves by banks and vegetation produce areas of calm water where communities develop. They are destroyed by vigorous wave action or a strong current. The members of these communities, small organisms living immediately below the surface and insects moving over it, can be observed and collected very easily from the banks. Although suitable habitats are common and widespread, the biology of communities living at the surface remains largely unexplored.

Animals and plants living at the surface are members of the community known as the neuston. Those which live above the water but in regular contact with it make up the epineuston, and those hanging down from the surface are the hyponeuston. The most impressive members of the epineuston are waterfowl, waterlilies and floating pondweeds. This handbook will be concerned, however, with the smaller, less striking invertebrate animals of the neuston. The most conspicuous of these are the arthropods. These include springtails, pondskaters, water boatmen, water beetles and water spiders, all of which will be familiar to a casual observer. Another familiar group, the flies, includes some species in which the adults cruise just above the water, or walk or skate across it, as well as many which lay their floating eggs on the surface to hatch out into enormously numerous larvae prominent in the hyponeuston. Other less conspicuous but no less interesting members of the hyponeuston include protozoa, rotifers and tiny crustaceans. Microscopic plants such as desmids and diatoms also abound, but this book confines itself to the fauna. Table 1 lists the groups of organisms common in the neuston.

All the conditions prevailing in the habitat relate to the boundary where air and water meet. Typically, it is rich in resources because materials collect at the surface, coming up from below or falling in from above. These materials may help to feed the community, or, sometimes, to destroy it. Oxygen is freely available from the air. The surface water undergoes greater fluctuations in temperature than either air or deeper water. Humidity, air and water temperatures and

Table 1. *Some of the organisms found in the neuston, and mentioned in the text (after Barnes, 1984, and Lee, Hutner & Bovee, 1985)*

Common name	Taxonomic group	Example
Kingdom Monera		
cyanobacteria (blue-green algae)		*Oscillatoria*
Kingdom Protista		
algae: diatoms, desmids, filamentous algae such as blanketweed, *green flagellates	'algae' includes many groups only distantly related to one another.	*Spirogyra*
protozoa (single-celled animals)		
	Phylum Sarcomastigophora	
*green flagellates	Subphylum Mastigophora	*Euglena*
amoebae	Subphylum Sarcodina	*Arcella*
ciliates	Phylum Ciliophora	
holotrichs		*Chilodonella*
peritrichs		*Vorticella*
spirotrichs		*Stylonychia*
hypotrichs		*Euplotes*
suctorians		*Podophrya*
metazoa (many-celled animals)	**Kingdom Animalia**	
rotifers	Phylum Rotifera	*Euchlanys*
flatworms	Phylum Platyhelminthes	*Polycelis*
snails	Phylum Mollusca	*Lymnaea*
arthropods		
crustaceans	Phylum Crustacea	
waterfleas	Order Cladocera	*Scapholeberis*
ostracods	Class Ostracoda	*Notodromas*
	Phylum Uniramia	
insects	Subphylum Hexapoda	
springtails	Order Collembola	*Podura*
water bugs	Order Heteroptera (=Hemiptera: Heteroptera)	
pondskaters (water striders)		*Gerris*
water boatmen (back swimmers)		*Notonecta*
lesser water boatmen (corixids)		*Corixa*
water crickets		*Velia*
water measurers		*Hydrometra*
beetles	Order Coleoptera	
whirligig beetles		*Gyrinus*
camphor beetles		*Dianous*
flies	Order Diptera	
mosquitoes		*Culex*
meniscus midges		*Dixella*
dolichopodid flies		*Poecilobothrus*
	Phylum Chelicerata	
arachnids	Class Arachnida	
spiders	Order Araneae	*Dolomedes*
water mites	Order Acariformes	*Atractides*

* Green flagellates may be classified either as algae or as protozoa. They are included as protozoa in the following pages.

exposure to sunlight will all contribute to changes in surface temperature, and there may be significant diurnal and other short-period variations. In winter, it may remain at a more or less constant 0°C. The ionic content of the very surface may not be the same as that of deeper water, which may possibly be significant for hyponeuston in brackish waters. Neuston will also be exposed to large amounts of solar radiation, especially ultraviolet wavelengths, which are potentially harmful. These may prevent some organisms from surviving at the surface, or may determine the times at which others are found there. Others may be protected by dark pigment from excessive exposure to ultraviolet radiation, which may explain the presence of melanin in the few crustacea which survive continuously at the surface. Behavioural responses to light intensity may be significant in all groups, particularly among surface-dwelling insects which are very mobile and able to move away from excessive light and heat. The restricted range of invertebrates (other than insects) occurring in the hyponeuston is perhaps related to the hazards of exposure to solar radiation.

The neuston is exposed to predators from above and below, as it is conspicuous at the surface. Small ditches lacking fish support dense populations of epineustonic insects, perhaps due to reduced predation.

True members of the neuston spend most or all of their lives at the surface, able to leave it only for brief periods or not at all. Other animals may visit it for shorter periods from above or below, such as the water spider *Argyroneta*, water boatmen (*Notonecta*) and large water beetles like *Dytiscus*. They may need to replenish air supplies or feed on surface prey, although they can capture prey very successfully under water and survive submerged for long periods. Animals collected from the surface may also include accidental visitors (such as leaf-hoppers and small spiders) which have fallen onto the water, perhaps from vegetation. They appear to be able to return to their preferred habitat by walking or running over the surface. Other transient visitors are emerging adult insects, including mayflies (Ephemeroptera), stoneflies (Plecoptera), caddisflies (Trichoptera), dragonflies (Odonata), and flies (Diptera) which have spent their larval lives in the water. Clouds of insects such as mayflies in spring and midges in summer and autumn can be seen escaping from the surface film. Practised observation will distinguish these temporary visitors from permanent inhabitants.

The neuston occupies a niche in which it is difficult to survive, and includes a limited range of organisms, requiring very specialised adaptations of structure and behaviour. How do they manage to remain at the surface when their tissues are universally more dense than water? What do they feed on? How do they escape the attentions of predators (or do they)? What do they do when their habitat becomes overcrowded, dried up or polluted? How do they survive the winter? In particular, we need more complete distribution records of surface-dwelling animals in Britain to assist in decisions about conservation, and we need a better understanding of the likely effects of pollution in this very susceptible habitat. This book aims to provide sufficient information to enable the reader to recognise and name the commoner animals present, and to learn something of the special nature of this community, in order to pursue further lines of investigation which may increase our understanding and interest.

2 The nature of surface forces

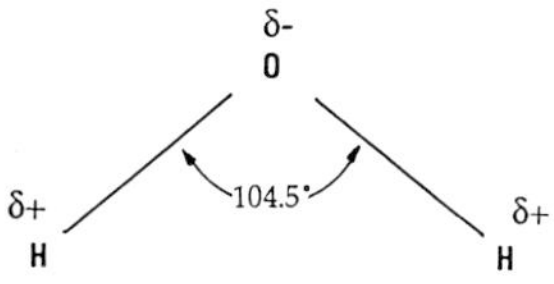

2.1 Surface tension

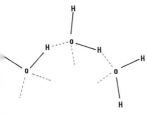

Fig. 1. The polarity of a water molecule. δ+ and δ- represent small local differences in charge.

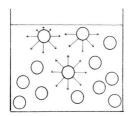

Fig. 2. Three water molecules, with hydrogen bonds shown as dotted lines.

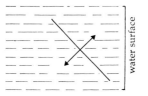

Fig. 3. The cohesive forces acting in a liquid.

Water is a unique substance with many properties that one would never suspect from the formula H_2O. The two hydrogen nuclei make an angle of 104.5° with the oxygen atom (fig. 1). The negative charge is increased in the oxygen part of the molecule, resulting in a partial positive charge on the hydrogen atoms. Such a molecule in which there is a distribution of charge producing negative and positive ends is said to be a 'polar' molecule. The positive pole of one molecule is attracted to the negative pole of a neighbour, giving rise to a directional bond which in water is known as a hydrogen bond. Although this is a weak bond, it is powerful enough to make water molecules highly attractive to each other (fig. 2). The mutual attraction of molecules of any substance is called cohesion. These polar forces give water high cohesion and help to produce its special liquid characteristics.

A molecule within a liquid is uniformly surrounded by others and subject to similar cohesive forces in all directions (fig. 3). For molecules at the surface between liquid and air, cohesive forces will act from below and to the sides, but not from above; the net force therefore tends to draw them back into the liquid. The liquid behaves in some respects as if it had an elastic skin. This effect is called surface tension, acting in the plane of the surface and tending to make its area as small as possible. Free-falling drops of water are spherical, having the smallest possible surface area for a given volume, due to surface tension. Because water has such high cohesion, it also has an unusually high surface tension. Only a few other pure liquids have a higher surface tension; liquid ammonia and mercury are examples.

Consider an imaginary line on the water surface, l m long. The surface tension on the right-hand side of the line pulls the line to the right with the force F; and an equal force to the left-hand side pulls it to the left (fig. 4). Surface tension, γ (the Greek letter, gamma), is the force per unit length acting on either side of any line drawn in the surface of a liquid. It is measured in milliNewtons per metre ($mN\ m^{-1}$). In older books, surface tension was given in dynes per centimetre ($1\ mN\ m^{-1} = 1\ dyne\ cm^{-1}$). Surface tension generally varies with temperature.

Fig. 4. The force due to surface tension acts at right angles to any line drawn on the surface of a liquid.

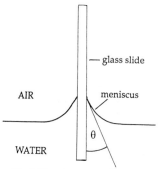

Fig. 5. The concave meniscus formed when a clean glass slide is placed vertically into water. θ (theta) is the contact angle.

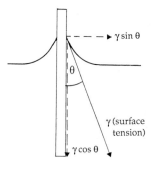

Fig. 6. The downward force due to surface tension operating in fig. 5.

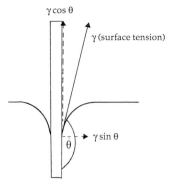

Fig. 7. The convex meniscus formed against a surface with low adhesion for water, placed vertically into water. The force due to surface tension tends to pull the object upwards.

Table 2. *Surface tension of water at different temperatures, and of mercury*

Temperature °C	Surface tension mN m^{-1} (=dynes cm^{-1})
Water	
100	58.9
50	67.9
20	72.7
0	75.6
Mercury	
20	435.5

Any molecule with atoms which are not symmetrically arranged will have some degree of charge separation. If this is considerable, the molecule is said to be 'polar'. A small molecule, or part of a molecule, with large charge separation will be very strongly polar. Because water molecules are strongly polar, they are attracted to molecules of other substances which have polar properties or which dissociate into charged ions. The attraction between water molecules and solid substances is called adhesion. The particular physical processes which make the interface between water and air such a specialised environment for organisms are surface tension and the adhesion of water for the surfaces of those organisms.

If a *clean* glass slide is placed vertically into water, the water rises up the glass to give a curved surface or meniscus (fig. 5). The water experiences an adhesive force to the glass surface and a cohesive force to itself. The balance between these two forces determines what happens to the water surface at the point where it meets the solid surface. Rarely will the water meet the surface at right angles. If adhesion is strong, the water will rise up the surface to form a concave meniscus. As the meniscus curves upwards, part of the surface tension force operates downwards (fig. 6). It tends to pull the slide into the water. The downwards force due to surface tension can be calculated by measuring the angle between a solid surface and the water meniscus at the point of contact. This angle is called the contact angle, θ (the Greek letter theta), and is always measured with respect to the liquid. The downward component of the force acting on the body will be:

$$\gamma \, l \cos \theta$$

where *l* is the length of the line of contact of the meniscus with the solid surface. The force will also have a horizontal component, proportional to sin θ. The notion of a contact angle gives a simple but effective way of expressing the adhesion between water and that solid surface.

If, however, a surface, such as part of an insect, has very low adhesion for water, the meniscus curves downwards (it is convex, fig. 7) and θ will be greater than 90°. Now the vertical component of the surface tension force acts upwards and tends to lift the insect out of the water.

Because clean water has such a high surface tension, that component can provide a force strong enough to support the weight of an insect with only a small part of its feet actually in the liquid, providing the line of contact. For epineustonic insects, the ability to stand on the water surface depends on a large contact angle on those parts of the insect surface in contact with the water. The body surfaces of epineustonic insects consist of materials with a low adhesion for water. Similarly an organism of the hyponeuston, such as a mosquito pupa (fig. 8), the weight of which is largely counterbalanced by buoyancy, only requires the perimeter of a very small low-adhesion area to hold it at the surface.

In general, the contact angle measures the degree of wetting of the most superficial layer of surface molecules of a solid. For a number of reasons (Adam, 1951*), it is very difficult to obtain consistent measurements with natural systems; contact angles are affected not only by contamination of a solid surface, but also by the surface roughness and by how long the surface has previously been in contact with water (see Holdgate, 1955 for further discussion). However, valuable information can still be obtained when investigating neuston simply by estimating the contact angle to the nearest 5° (techniques, p. 77). This information helps us to gauge the relative 'wettability' of body surfaces. We can divide substances into hydrophilic (water-loving) and hydrophobic (water-hating) kinds, depending on the degree of adhesion which they show for water. Hydrophilic surfaces attract water and are easily wetted, and exhibit small contact angles. Hydrophobic surfaces are said to repel water; they are not easily wetted. In fact, the universal attraction between molecules of all substances ensures that the adhesion of a surface for water is never zero; hence in real systems the contact angle is never as much as 180°. The much greater cohesive force of the water molecules themselves will cause water in contact with hydrophobic surfaces to round up into spherical droplets, as when rain runs off petals and leaves of plants. Here the contact angle is high, 100° or more. Loosely speaking, an 'unwettable' surface is one with a contact angle to water greater than 90°, and a 'wettable' surface is one with a contact angle less than 90°.

One important consequence of having an unwettable surface is that if animals come up from the depths to break the surface, to leave the water or to take in fresh air supplies, the surface tension of the water pulls the water away from any hydrophobic area – literally 'like water off a duck's back'!

2.2 Surface films on natural bodies of water

Natural bodies of water will rarely, if ever, consist of clean pure water. The surface will be contaminated with

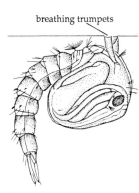

breathing trumpets

Fig. 8. A mosquito pupa, hanging from the surface by the adhesion of a small area of contact.

*References cited under the authors' names in the text appear in full in Further Reading on p. 79.

material such as dust and debris falling onto it, the films of molecules shed by any living or dead objects which penetrate the surface (such as spores, leaves, or animal bodies), and the substances pushed out from the water. These will include the secretions and excretions of living organisms and the products of their decay, such as proteins, polysaccharides, glycoproteins and lipids, some of which will reduce the surface tension and consequently the forces on which many members of the neuston depend. Compounds which have this effect are called surfactants (surface-active agents or 'tensides'). These molecules generally consist of one part, usually referred to as a polar group, which has a high adhesion for water, either because of its polar asymmetry or because it is ionised and carries a charge. The other part has a low adhesion for water because it has little or no charge separation. Many organic molecules consist of a 'backbone' of long hydrocarbon chains, or, less commonly, benzene rings, having little or no charge separation and hence low adhesion for water, combined with a variety of polar side-chains which are strongly attracted to water such as the amino (-NH_2), carbonyl (-CO), hydroxyl (-OH) and carboxyl (-COOH) groups. These groups will adhere strongly to the surface water, whilst the hydrophobic part will lie on the surface (fig. 9), as there is little force to pull it back in. The surface is covered by a single layer of molecules which reduce the surface tension. Modern industrial and domestic detergents work on the same principle, and are designed to reduce surface tension sometimes by half or more. Together with soaps, they are the most familiar surfactants. Soaps are usually the sodium or potassium salts of long-chain fatty acids which ionise in water, like ionic detergents, to give charged particles. Non-ionic detergents have hydrocarbon chains attached to groups such as hydroxyl (-OH) or bisulphite (-HSO_3) which attract water. The presence of detergents will obviously have considerable effects on neustonic animals, which will be considered further below. Simple techniques for investigating the effects of surfactants on the surface tension of natural waters are described in chapter 7. Modern theories of detergence are discussed in many books on surface chemistry such as Aveyard & Haydon (1973).

Goldacre (1949) reported that many natural bodies of water have a film which behaves like a monomolecular layer of protein or lipoprotein. Apart from this rather basic account, we have little information about natural surfactants on freshwater. More is known about the sea, where organic films produce the familiar slicks over Sargasso weed, which 'calm' the surface. Sturdy & Fischer (1966) showed that, downwind from Kelp (seaweed) beds, the surface tension was reduced from 74 mN m^{-1} to 50 mN m^{-1}. Baudoin (1955) made many measurements of surface tension from marine and freshwater environments. Dissolved salts increase surface tension by producing charged ions which are

Single surfactant molecule

—Hydrocarbon chain

Polar group

Fig. 9. Surfactant molecules forming a monomolecular layer on the water surface.

strongly attracted into the water. Baudoin gives a maximum value for seawater of 74.6 mN m^{-1}, compared with 72.25 mN m^{-1} for pure water. The foam produced by wave action on sandy beaches gave lowered values similar to those of protein solutions. Blotting with tissue raised the surface tension temporarily almost to its normal value, but it fell again to 55 mN m^{-1}. He was able to repeat this process four times and suggested that the surface-active film was replaced from the water mass. Baudoin described the foam as a death trap for small insects such as dolichopodid flies, which are no longer supported by the film and drown. Surfactants from water are concentrated in foam; liquid collected as foam is a much richer solution than the water body from which it is collected (Barger & Garrett, 1970). Many surfactants in solution accumulate so quickly at the interface that it is surprising that Baudoin could measure the increased surface tension after he had blotted one lot off the surface and before another lot took its place.

MacIntyre (1974) described a 'microlayer' in the top millimetre of the ocean, consisting of the surface film and the water just beneath it, which is quite different in chemistry from the rest of the water. It contains surfactants of two kinds, both of which may originate from the plankton. 'Dry' surfactants are 'fatty' compounds consisting largely of strongly hydrophobic hydrocarbon chains sticking out of the water, anchored by small hydrophilic groups. They include hydrocarbons, sterols, fatty acids, glycerides and phospholipids. 'Wet' surfactants contain a predominance of hydrophilic groups with a few hydrophobic side-chains anchoring them to the surface. They are generally polysaccharide-protein complexes; glycoproteins and proteoglycans are a major part of surface films in the sea (Sieburth, 1976). Bubbles will collect surfactants at the gas–water interface which surrounds the bubble and, on breaking at the surface, will enrich the microlayer.

Bursting bubbles may also deposit solid particles including bacteria at the surface. Organic particles are used as food by bacteria and zooplankton. In the sea, slicks may contain bacterial populations a hundred times as dense as those in deeper water (Crow and others, 1975). Johnson (1976) found that the highest concentrations of non-living organic particles occurred during the spring plankton outburst and were correlated with diatom growth and windy weather. Wind increases wave action and bubble formation. Similar processes probably occur in freshwater, where bubbles may arise also from photosynthesis, or decomposition in anaerobic mud. Another source of surfactants here are the humic substances derived from lignin, cellulose and plant proteins, washed in from the soil.

The organic film on freshwater bodies shows very variable development at different times. It may be ingested by neuston or decomposed by bacteria. Breaking bubbles will throw fragments into the air as aerosols. The film is also degraded by visible and ultraviolet light (Baier, 1972). It may be dispersed by strong winds or heavy rain. Wind, water currents and wave action can pile up dry surfactants, together with other floating objects, in sheltered backwaters. Floating weeds, dead leaves and other debris provide a platform where less specialised neuston can live. Variations in the nature of the surface determine the type of community present.

A comprehensive review of papers on the surface film is given in Wangersky (1976).

3 How animals exploit surface forces

3.1 Support

The boundary between air and water is like a strong, resilient membrane. Many kinds of arthropod exploit its surface properties to stand, walk or glide, or use it as a springy trampoline.

A body at the surface is supported by its buoyancy and by surface tension. The ability of an animal to be supported depends on its effective weight; its total perimeter of contact (l) with the water; the surface tension of the water (γ); and the contact angle (θ) between water and the body surface.

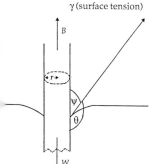

γ (surface tension)

B

r

ψ
θ

W

Consider the simple case of an insect's foot, of cylindrical shape, resting vertically in the water surface (fig. 10). The surface of the foot is water-repellent (θ is more than 90°) producing a convex meniscus. In this case, the length, l, of the line of contact of the meniscus with the insect's surface equals the circumference of the foot, $2\pi r$, where r is the radius of the foot. The upward force due to surface tension acting around the cylinder is $2\pi r\gamma$, acting at an angle ψ (the Greek letter psi) to the vertical. If F is the component of the force acting vertically upwards,

$$F = 2\,\pi\,r\,\gamma\cos\psi.$$

Buoyancy (B) is the upthrust due to displacement of water by any part of the body situated below the water surface.

The forces acting upwards, F and B, are counteracted by the downward force of the weight, W, of the insect.

$$W = mg$$

where m is the body mass, and g is the acceleration due to gravity.

If $W = F+B$, the animal is supported.

Beament (unpublished information) gives, as a rule of thumb, a required perimeter of contact of about 0.3 mm mg^{-1} of insect, where the contact angle is 120° and the water clean. A 5 mg insect would therefore be supported by six hydrophobic legs, each with a diameter of only about 0.1 mm at the level of contact.

Clearly, there must be an upper limit to the mass of an animal living on the surface. Larger insects of the epineuston such as pondskaters, *Gerris*, water measurers, *Hydrometra* and water crickets, *Velia*, have lightweight bodies and long legs, separating the 'dimples' formed beneath each foot, so that each dimple makes its own effective contribution to the forces supporting the body. In the slender pondskaters, the line of contact l is increased by the tibiae as well as the feet (tarsi) of the third pair of legs

Fig. 10. The forces affecting an insect's foot in the water surface.

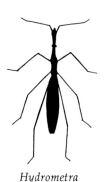

Hydrometra

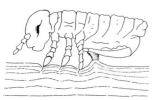

Fig. 11. The springtail
Podura standing on the water
surface (after
Wigglesworth, 1964).

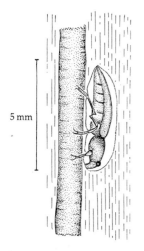

5 mm

Fig. 12. A beetle, enclosed in
an air bubble, crawling down a
stem (after Thorpe, 1950).

Scapholeberis

lying on the water surface. The tibiae of the hind legs and the tarsi of all six legs carry a dense covering of unwettable hairs on the regions which come into contact with the water. Baudoin (1955) observed that *Gerris* was still supported without falling through or getting wet when the surface tension was lowered by 10%, which indicated a margin of safety enabling it to survive even in habitats where mild pollution has reduced the surface tension.

A typical specimen of *Gerris lacustris* weighs 15 mg, and many epineustonic insects are much smaller, such as flies and the bugs *Hebrus* and *Microvelia*. As mass gets smaller, the need for large perimeters of contact is reduced, so that the smallest insects such as the springtail *Podura* (fig. 11) balance on 'points'. Tiny beetles which live amongst aquatic vegetation are so unwettable that they can enter the water only by crawling down plant stems, enclosed in an air bubble which they use for breathing. Only the tips of the feet, mouthparts and abdomen emerge from the bubble into the water (fig. 12).

In the hyponeuston, effective mass may be reduced by buoyancy devices such as the oil droplets of diatoms and the gas bubbles of *Arcella*. The air contained in the tracheal (respiratory) system of pupal mosquitoes is important in providing the buoyancy which maintains them just below the water surface.

Most invertebrates living in freshwater have bodies which are easily wetted. Many, including protozoa, rotifers, crustacea, flatworms and many insect larvae, have a relatively small mass, and may be regular or transient members of the hyponeuston. Support due to surface forces may be aided by hydrophobic bristles or hairs. The specialised crustacea of the hyponeuston, *Scapholeberis* and *Notodromas*, depend on rows of bristles (setae) to suspend them. These are elaborately developed in *Scapholeberis*. The ostracod, *Notodromas*, makes a little hump in the water as it swims just below the surface. Are the fringes of hairs seen in creatures such as the nymph of the water mite, *Atractides*, also able to support them using surface forces?

Conversely, the strength of adhesion is shown by the vain struggles of small animals which fall onto the water and cannot escape. Waterfleas such as *Chydorus* which normally live below the surface appear to experience a similar problem if they swim up to the surface. The contact angles of various crustacean cuticles might be a fruitful field of investigation.

3.2 Arthropod cuticle

Many arthropods, particularly insects and spiders, are coated with a thin waxy layer which protects their bodies from water loss by evaporation in air. The outermost layers of an insect are generally waterproofed with wax and coated with cement. The cement layer varies in thickness and extent, and is lacking in many insects. Locke (1974) suggests that, like varnish, it forms a hard transparent outer layer with a protective function. In some insects, waxes are also deposited as 'blooms' on top of the cement, in thin sheets or plates, which offer protection against submergence for some small insects. Not all insect cuticles are uniformly water-repellent; some have wettable surfaces which Locke considers may be due to hydrophilic cement overlying the waxes.

The chemistry of cuticle has been much investigated and there are several comprehensive reviews (Hackman, 1974; Locke, 1974; Beament, 1976; Gilby, 1980; Hadley, 1981). The composition of the waxes varies greatly in different species, but hydrocarbons and free fatty acids are generally present. The former make up more than half the total lipid and are usually alkanes and alkenes of chain length C_{20}–C_{50} (Gilby, 1980). Such non-polar compounds make the cuticle strongly hydrophobic and waterproof, as shown by the high contact angles for water of many insect bodies. This relationship was first investigated by Brocher (1910) in an extensive study on which much later work is based. In spite of the difficulties of measuring contact angles, Holdgate (1955) devised a way to assess the relative wettability of various insect cuticles. High contact angles indicate strongly water-repellent cuticles. They are found in insects such as the water beetle, *Dytiscus*, the water boatman, *Notonecta* and the whirligig beetle, *Gyrinus*. *Dytiscus* and *Notonecta* can submerge and swim underwater for long periods, but they surface periodically to breathe. *Gyrinus* lives on the surface. All of them can leave the water at will and take to flight, aided by the smooth hydrophobic surfaces produced by wax and cement layers. When a water beetle emerges from the water, droplets form on the back which are easily thrown off. In the whirligig beetle, the cuticle in the lower half of the body is feebly attracted to water, having a lower contact angle than the water-repellent upper half (Holdgate, 1955). The body floats at the same level whether the beetle surfaces from below or above (fig. 13). The beetle can submerge, overcoming the surface forces with its strong swimming limbs. When it approaches the surface from below, it bobs up and floats. Its water-repellent cuticle also greatly reduces the muscular effort needed to escape from the film and become airborne.

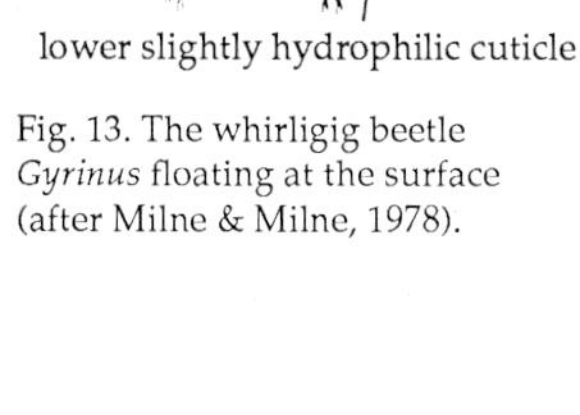

upper, water repellent cuticle

upper part of eye

water surface

lower part eye

lower slightly hydrophilic cuticle

Fig. 13. The whirligig beetle *Gyrinus* floating at the surface (after Milne & Milne, 1978).

3.3 Hair piles

In arthropods, hydrophobic surfaces are produced when a layer of closely-packed short hairs are made unwettable by waxy secretions. The layer traps a cushion of air. Water sprinkled onto velvet collects in spherical drops. The water has a very high contact angle in relation to the surface of the pile. The drops look silvery because of the air trapped beneath them, causing reflection of some light from

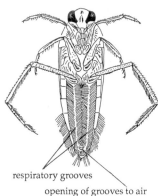

respiratory grooves

opening of grooves to air

Fig. 14. The water boatman
Notonecta, to show the
respiratory grooves covered
by unwettable hairs (after
Portier, 1911).

tracheal system: the
respiratory system of an
insect; a system of air-
filled tubes

spiracles: the openings of
the tracheal system at the
surface of the insect's body

the water–air boundary. The silvery appearance of many animals seen under water is due to retention of a layer of air over all or part of the body. It is held by a velvet-like pile of unwettable hairs set so close together that water cannot penetrate between them. The pile will remain completely dry. The sturdily built swamp spider, *Dolomedes fimbriatus*, typically weighs about 28 mg, yet it can scuttle across the open water of the narrow drainage ditches where it lives without apparently breaking the surface. If it is startled, it submerges quickly and is able to remain submerged with the whole body covered by a silvery film of air. Like other animals, such as pondskaters, with a strongly hydrophobic hair pile covering the body, it can escape from the surface only by climbing down emergent vegetation. Many terrestrial arthropods have a dense covering of unwettable hairs which enable them to avoid a watery grave by walking on the water without sinking.

Unwettable hair piles ensure an air supply for many neustonic animals when they are submerged. The pile may cover the whole body, as in *Dolomedes*, or may be restricted to certain areas as in *Notonecta*, the water boatman. Here, air is trapped in two long grooves, one on each side of the abdomen, and in the hairs of the ventral surface (fig. 14). As with other insects which carry down an air bubble, *Notonecta* breathes air through its spiracles, like a terrestrial insect. The bubble acts as a compressible physical gill. As oxygen is removed by breathing, more will diffuse in from the surrounding water. The length of time during which the insect can remain under water depends on the amount of oxygen dissolved in the water. In winter, when low temperatures increase the solubility of oxygen and decrease the animal's metabolic rate, it can survive continuously submerged. In summer, when it is very active and the oxygen content of the water falls as the temperature rises, it must surface to replenish its air supply at frequent intervals. Cockrell (1984) found that, between August and October, it is generally found floating freely on the open water surface. At rest, it hangs upside down from the surface with the tips of the feet, and the tip of the abdomen, projecting slightly above the surface. When it stops swimming, its bubbles ensure that it bobs up to the surface. The buoyancy of its air bubbles supports it, as well as providing a safety mechanism against drowning.

Impermeability of the cuticle to gases is the price that must be paid by arthropods for a cuticle that is waterproof. The springtail, *Podura*, has a cuticle covered with minute bumps having hydrophobic tips, making the outer body surface unwettable and preventing the animal from becoming trapped in the surface film. Its wettable ventral tube anchors it and prevents it from being blown across the surface. *Podura* is an atypical insect in that it lacks a tracheal system. It respires through its cuticle. The spaces between the minute bumps with which it is covered are lined with hydrophilic cuticle, which is permeable to oxygen (Noble-Nesbitt, 1963a).

Many aquatic insects groom themselves regularly. Grooming ensures that the hairs of a hair pile are evenly disposed to prevent water penetrating between them, and may maintain the hydrophobic properties essential for support and breathing, by spreading fresh lipid to replace any lost from hydrophobic areas.

boat shape cut from balsa wood

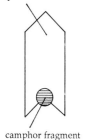

camphor fragment

Fig. 15. A camphor boat.

Velia

3.4 'Expansion swimmers'

A few insects, like the staphylinid 'camphor' beetles, *Stenus* and *Dianous* (pl. 2.8), secrete a mixture of powerful surfactant chemicals, the most important of which is stenusin, a camphor-like terpenoid substance. The secretion spreads from the tip of the abdomen as a monomolecular layer behind the beetle, lowering the surface tension, so that the insect is pulled forwards by the greater surface tension of the clean water in front. The beetles can reach speeds of 45–70 cm sec^{-1}, about 25–35 times as fast as they can run (Nachtigall, 1974). These 'jet-propelled' beetles are found at the edges of waterfalls on mountain streams, clinging to moss as they lie in wait for springtails. Their great speed helps them in catching such lively prey. Toy camphor boats work on the same principle. A boat-shaped piece of balsa wood sheet with a notch at one end containing a piece of camphor (fig. 15) will shoot along the surface when placed on water. A similar toy using detergent instead of camphor is described by Ward (1985).

The bug, *Velia caprai*, can move in similar fashion, supplementing its normal walking movements. It squirts a surfactant secretion backwards from its proboscis. Linsenmaier & Jander (1963) allowed *Velia* to walk on a water surface sprinkled with *Lycopodium* powder. The surfactant secretion pushed the powder aside leaving a clear trail behind the insect. Talc may be used as a cheap substitute for *Lycopodium* powder, to try this out with other insects (techniques, p.76).

4 The surface community

4.1 Food sources

A still water surface accumulates material which floats: animals, pollen, leaves, seeds and organic molecules with surfactant properties. This material may be eaten directly by neustonic animals, or broken down further by the bacteria and fungi of the decomposer food chain. These micro-organisms, by building the available chemical energy into their bodies, concentrate resources into the form of particles which can be ingested more easily by small-particle feeders such as protozoa, rotifers and crustaceans. The micro-organisms which derive their energy directly from the chemical substances in the film, together with the micro-organisms which feed on them, may be called the microneuston. In the sea, at least, this heterotrophic community may develop populations, particularly of bacteria and protozoa, as dense as may be obtained in optimum conditions for growth in laboratory cultures (Sieburth and others, 1976).

Another major food source is the algae which as primary producers form the basis of the grazing food web. Planktonic algae may collect just beneath the surface either actively, in response to the stimulus of light, or passively, due to flotation devices such as oil droplets or gas vesicles.

The microneuston and the algae form the basis of the neustonic food chain. Very large populations of fly larvae, and of waterfleas, feed by filtering out particles, living or dead, from the surface. The total mass of midge or mosquito larvae, in particular, may be very large. They provide a large part of the menu of many of the neustonic predators. Yet knowledge of feeding habits and relationships is incomplete. Gut contents of some species could be studied. The remains of algae, crustacea and insects can be identified with practice. Bugs and flies suck their prey and have no identifiable contents in the gut, but the prey seen on the proboscis can be recorded. Shrivelled sucked remains of mosquito and midge larvae can be recognised on the surface or in laboratory cultures. Observations on natural predation in the field are most valuable.

Matthey (1971) studied the semi-aquatic insects of peat-bogs for several years. The predator food chain was based on chironomid larvae, springtails, hydrophilid beetles and frog tadpoles. During July and August, large numbers of jassid bugs (especially *Macrosteles sexnotatus*) fell onto the water from surrounding vegetation. These plant bugs were an important part of the predators' diet, replacing the springtail, *Podura*, and chironomids which had decreased or disappeared by this time. *Podura* was a major food supply for pondskaters, *Gerris*, water crickets, *Velia*, the beetle,

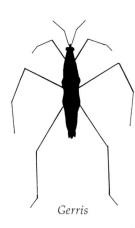

Gerris

Hydroporus, and the flies, *Ephydra* and *Hydrellia*. Large springtail populations were always accompanied by the bug *Hebrus ruficeps* and the first two nymphal stages of several *Gerris* species, for which the springtails provided food. The springtails, and *Hebrus* and *Gerris*, can all overwinter in the sphagnum. Populations of six species of *Gerris* were observed. They could survive on three basic food sources: *Podura*, *Macrosteles* and chironomids. Adult chironomids were eaten by *Gerris*, *Velia*, dolichopodids and spiders. The swamp spider, *Dolomedes fimbriatus*, captured tadpoles, small frogs and newts.

In Matthey's study, the larger the free water surface, the more varied was the neuston community observed. Gerrids were widespread, some species preferring the running water of the canals, others preferring ponds. As the ponds dried up, it became difficult for the gerrids to move on the mud, and they tended to migrate onto the canals, or became torpid in the sphagnum. Their removal favoured smaller forms like *Podura*, *Hebrus* and the dolichopodids, which became much more numerous. There were practically no predators catching *Gerris*, but their numbers were kept low by intense cannibalism. The fifth nymphal stage and the adults preyed heavily on the young nymphal stages.

Transient visitors to the neuston include pond snails and flatworms (pl. 8). They hang upside down by their flattened ventral surfaces, or glide along beneath the surface. They suck up the film, binding together the useful particles with the mucus which they secrete freely. Neuston supplements their other food sources, such as floating vegetation, a major food supply for aquatic snails.

Neuston is a potential food source for aquatic, semi-aquatic and terrestrial predators. The fish, of course, could be regarded as the top predator in the neuston ecosystem, and neustonic insects must contribute to the diet of many freshwater fish. Brown trout, *Salmo trutta*, are reported to feed on the bug, *Velia caprai*, in the field (Bronmark and others, 1984), although these authors report that the fish find the bugs distasteful and release them, still alive, after capture; only a few bugs are swallowed. The occurrence of schools of prey may improve their chances of survival by improving the rate at which the predator learns to avoid prey. Thanatosis (shamming dead) is seen in several species of water bugs as an anti-predator device. Whirligig beetles are known to produce a secretion noxious to fish (Benfield, 1972) and beetles are ejected by predatory fish in an undamaged state. Their schooling behaviour and general structural toughness, which they share with *Velia*, may similarly aid in their survival, as Bronmark and others suggest.

4.2 Microneuston – protozoa, rotifers and crustacea

Some organisms of the hyponeuston hang from the air–water interface. Others attach themselves to objects floating in the surface. Many aquatic micro-organisms need solid surfaces on which to grow. Accumulations of decaying flotsam include decomposing remains such as skeletonised leaves which may provide the basis on which a community can develop.

The wealth of this community can be illustrated by considering a particular example. A rich source of microscopic hyponeuston was found by the author after a period of fine calm autumn weather, when shallow ponds produced a thick green crust on the downwind surfaces. Ducks swimming through it left a clear wake which was gradually filled in again. The basis of the deposit appeared to be the cuticles or decayed epidermises of leaves, consisting of brown fragments of flattened film which could be lifted from the water and placed flat on slides (fig. 16). Looking like irregular wire netting under the microscope, the meshwork provided a scaffolding for the accumulation of a rich variety of organisms. The brown colour was due to many species of attached diatoms. Moving freely among the leaf remains were enormous numbers of a large *Euglena*, and fewer but very numerous dinoflagellates, *Glenodinium cinctum*. The green colour of the crust was produced by the dense aggregates of *Euglena*. The colonial flagellates *Synura uvella* and *Anthophysa vegetans* were also abundant. None of these flagellates is purely neustonic but they are often found at the surface during favourable still conditions.

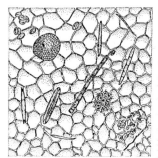

Fig. 16. Microneuston living on the surface of a floating leaf skeleton (from photograph).

Among the single-celled organisms, the Protista, common hyponeustonic forms include the floating suctorian, *Sphaerophrya*, and the ubiquitous shelled amoeba, *Arcella*. In young individuals of *Arcella*, the beret-shaped shell or test was still a pale straw colour. In old individuals, it had become a deep orangey brown owing to the deposition of iron and manganese compounds. *Arcella* usually hangs upside down suspended by its few pseudopodia, its buoyancy sometimes increased by tiny gas bubbles in the cytoplasm. These specimens had the cytoplasm densely packed with food vacuoles containing diatoms and green algae. There were also many monopodial (single-footed) amoebae. These occur regularly in the hyponeuston, but are very difficult to identify. Attached forms include the ciliates, *Vorticella*, *Tintinnidium* and *Stentor*, and also the suctorian *Podophrya*, a predator feeding on the numerous free-swimming ciliates, particularly *Euplotes* and *Coleps*. Most of the ciliates found were filter-feeders. The rich supply of bacteria and tiny Protista explains the occurrence and multiplication of these and other small-particle feeders, including free-swimming rotifers such as *Squatinella rostrum* and *Euchlanys* species, and the attached rotifers *Rotaria neptunia* and *Collotheca* species. Some of these are more typically found in the plankton; their abundance at the surface was presumably due to the very abundant food. Other rotifers which have been found in similar conditions are *Keratella* and *Brachionus*.

Podophrya

Some of the animals feeding at the surface bear epizoites. These are small organisms, generally Protista, living on the surface of larger organisms, particularly Crustacea. Stalked forms such as the suctorian *Ephelota* and the peritrichs *Vorticella* and *Zoothamnium* attach themselves to a mobile host and profit from the rich food supply when the host itself is feeding at the surface film.

Rotifers are commonly found in surface collections. Most of those found are females, which lay eggs without being fertilised, that is, they reproduce parthenogenetically. At certain times, the same species of rotifer may go through a sexual phase, with both males and females. Several parthenogenetic female generations derived from a single egg can achieve rapid colonisation of small bodies of water, so that sudden outbursts of a single species are typical. The sexual phase produces resistant eggs which are distributed widely by wind and by animals. Many genera have a worldwide distribution and live in a wide range of conditions. The alternation of sexual and parthenogenetic phases leads to great variability of form which makes some rotifers difficult to classify.

In large bodies of water like lakes, rotifers appear to avoid the zone immediately below the surface, although they reach maximum numbers in the upper water layers. Active vertical diurnal migration occurs; they are highly sensitive to light. In smaller habitats, rotifers are frequently found at the surface. Hyman (1951) refers to three genera as neustonic: *Beauchampia*, *Ptygura*, and *Collotheca*. The latter two have gelatinous envelopes which help them to float. All three attach themselves to floating objects, preferring the finely divided leaves of plants like bladderwort, *Utricularia*, which is a neustonic insectivorous plant. Hyman states that different species of rotifer select different species of plant such as the cyanobacterium *Glœotrichia*. Cyanobacteria have gas vesicles which make them buoyant. They can regulate their buoyancy in relation to light intensity (Reynolds & Walsby, 1975). Vertical movements of rotifers may be linked with those of algal blooms. Semi-planktonic species capable of attachment by their 'feet' accumulate amongst planktonic algae and rapidly produce swarms. The 'blooms' typical of cyanobacteria create ideal conditions for attached microneuston, as well as a habitat for crawling waterfleas such as *Chydorus*.

There is much to be discovered about the occurrence and habits of rotifers at the surface of small bodies of water. They turn up in large numbers even in small garden ponds.

The Cladocera or waterfleas are small crustaceans which use limbs with combs of bristles to create a feeding current from which they sieve food particles. Cladocera are abundant in the plankton and amongst floating weed. Their role in the neuston needs further investigation. The common genus *Chydorus* is frequently collected at the surface. More specialised for life in the neuston is *Scapholeberis*. Scourfield (1900) first described it as 'hanging from the ceiling of the

Scapholeberis

pond' where it draws the nutritious film into its mouth. It is suspended upside down by the rims of the shell valves, which are shaped like runners, and which lie in the plane of the surface. The uppermost (ventral) surface is much darker than the lower (dorsal) surface, because of the presence of the dark pigment, melanin (pl. 6.3). This inverse colouration may protect the waterflea, perhaps by making it more difficult for predators to see it from above as it merges with its muddy background, or by intercepting harmful ultraviolet radiation. Another small crustacean, the ostracod *Notodromas* (pls. 6.1, 6.2) is found in claypits and shallow ponds. It shows similar modifications of structure and colouration for neustonic life, resembling its African relative, *Oncocypris mülleri* (Fryer, 1956).

Planktonic crustacea multiply rapidly and become very abundant when conditions are favourable. Their daily vertical migrations bring them close to the surface at certain times, when they provide a rich food source for small predators like the bugs *Hebrus* and *Microvelia* and small flies.

4.3 Predation

The water surface is like a spider's web which traps many animals which fall onto it and cannot escape. The floating corpses of aphids, thrips, flies and other insects are sometimes very abundant and provide a major food source for predators. Many flying insects are attracted to water by the light reflected from the surface. In the spring, winged pondskaters are strongly attracted to light-reflecting surfaces such as ponds or even car bonnets and roofs. Presumably they are seeking new habitats (Landin & Vepsäläinen, 1977). On the other hand, many terrestrial insects dive to watery graves, perhaps disorientated by reflected light.

The air–water interface propagates the slightest ripples caused either by predators themselves or by insects that have fallen in. There is no doubt that many predators depend on detection of surface waves in hunting prey. The spider *Pirata piraticus* spreads its feet widely and can locate prey by responding to the amplitude of ripples passing under the feet at different distances from the source (Savory, 1964). Land spiders respond in the same way to prey vibrating their webs. The swamp spider, *Dolomedes fimbriatus,* sits on the projecting stems of plants in the narrow dykes and ditches where it often lives, resting its front legs on the water. It responds immediately to an insect dropped up to 40 cm away, by turning quickly and running towards the prey. Bleckmann & Rovner (1984) found that single-frequency signals of 40–70 Hz produced the sharpest response. These authors tested the role of the eyes in hunting as the spiders have an impressive battery of eight eyes. They found that blinded spiders showed a decrease in both running distance and running speed, but they

considered that vision is not needed for prey capture in
several species of *Dolomedes*.
        The secondary role of vision in prey capture for
some predators is confirmed in *Notonecta*. Both *N. glauca*
and *N. obliqua* eat more in the dark (Giller & McNeil, 1981).
The ability of *N. glauca* to react to surface vibrations is well
known. Sense organs which respond to mechanical
stretching (mechanoreceptors) are situated in the feet and in
the joints of all three pairs of legs. There are also rows of
sensory hairs on the last three abdominal segments, in
contact with the water surface when the animal is at rest

(fig. 17). Lang (1980) considers that these two sets of sense
organs can discriminate between disturbances set up by
struggling prey on the water and waves set up by
Notonecta or by non-living objects. Low frequencies, up to
about 50 Hz, stimulate the mechanoreceptors on the legs,
whilst the abdominal receptors respond best to higher
frequencies (50–300 Hz). Stimulation of both systems is
necessary to set up prey-catching behaviour. Prey-waves are
higher frequency waves which elicit the capture of prey,
including first-stage nymphs of the insect. This explains
why young nymphs are cannibalized. Older nymphs and
adults produce wave signals at frequencies below 40 Hz and
are not attacked. 'Prey signals' include frequencies between
70 and 140 Hz.

Fig. 17. The water boatman
Notonecta resting below the
surface.

        Murphey & Mendenhall (1973) showed that the
American species, *N. undulata*, can orientate to prey almost
perfectly. If the mechanoreceptors on the legs and abdomen
are removed, but the eyes are intact, the ability to orientate
to prey is impaired.
        Prey-waves are set up by the leg movements of
struggling insects trapped in the film, according to Lang
(1980). He also found that the waves generated by surfacing
insects produced a predatory response. Vibrations due to
swimming were of less importance. Larval culicine
mosquitoes, rather than anopheline mosquitoes, were the
preferred prey of all *Notonecta* instars, and were captured
when they surfaced to breathe or dived again. Corixids
(lesser water boatmen) were caught in the same way, and
also mayfly nymphs coming to the surface to emerge.
Notonecta could detect *Gerris lacustris* more than 14 cm away
and catch it. This confirms the view of Scott & Murdoch
(1983). They regarded it as the top predator in stock tanks in
California, where *N. hoffmanni* was reared. Its predatory
behaviour is stereotyped and inflexible. Size of prey is
important. Nymphs feed voraciously on waterfleas, while
adults seem to prefer mosquito larvae.
        The behaviour of *Notonecta* allows it to make the best
use of seasonal food supplies. In the summer, it spends most
of its time at the surface, waiting to respond immediately to
any disturbance due to prey. There is an abundance of
trapped and emerging insects on the water, as well as fly
larvae and waterfleas in the hyponeuston. In winter, when
there is very little neuston, it hunts on the bottom.

The pondskater, *Gerris*, also detects surface waves, using sense organs on the flexible membranes between the segments of the feet and middle legs. They monitor the surface for vital information about potential prey and predators. Skaters drum on the surface as part of their mating and territorial behaviour. In the Australian species of *Rhagadotarsus*, males produce large amplitude waves, using the middle and possibly also the hind legs. This appears to attract females (Wilcox, 1972). Males and females both set up low amplitude waves by beating their forelegs synchronously on the surface. Mating can occur in total darkness. Aggressive signals are also used by males to defend mating sites. Fights lasting several minutes may occur between males.

Surface ripples are also used by insects which live in, or just below, the surface. Whirligig beetles are largely carnivorous, having powerful biting mouthparts. They appear to distinguish prey waves from the ripples they create themselves. Gyrinids are active at night when visual signals would be useless. *Orectochilus*, the hairy whirligig, is rarely seen on the surface in the daytime (Balfour-Browne, 1950). Other beetles, which hunt in the water, may sometimes feed at the surface. The great diving beetle, *Dytiscus*, takes moths which fall onto the water at night.

Various larvae and pupae in the hyponeuston react swiftly to disturbance. Presumably prey organisms as well as predators monitor surface vibrations. Some of them must escape, as material for natural selection!

4.4 Water bugs

Water bugs, including the pondskaters, water crickets, water measurers and water boatmen (table 1, p.2) are identified as bugs by their sharp beak. Generally using the front pair of legs to hold the prey, they use the beak to pierce their prey and suck its juices. Both *Gerris* and *Notonecta* secrete toxic saliva which rapidly subdues prey and digests the body contents. They may both inflict a nasty bite if handled and *Notonecta* may be a nuisance in open-air swimming pools. Both are cannibalistic, attacking younger members of their own species which are slower and vulnerable because of their softer cuticle. Such attacks are common in the early part of the season when other food sources may be scarce. Pondskaters tend to collect in groups and will band together to attack large prey such as a wasp. Water boatmen are fiercer than pondskaters with which they compete for prey, and hunt singly in the water as well as at its surface. Cockrell (1984) observed that, in summer, *N. glauca* feeds on the abundance of trapped and emerging insects present at the surface, together with the culicine mosquito larvae and waterfleas which disturb the surface. In the winter, it remains submerged, feeding on bottom-living invertebrates. *Notonecta* has a voracious appetite for

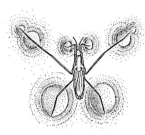

Fig. 18. When the pondskater *Gerris* stands on the water, its feet and legs 'dimple' the surface (from a photograph by D.M. Guthrie).

Arrows indicate magnitude and direction of corresponding body velocities

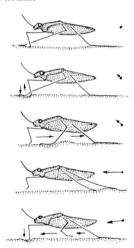

Fig. 19. Side view of the pondskater *Gerris* during a glide ahead (after Darnhofer-Demar, 1969).

its preferred diet of mosquito larvae. Water bugs might well be used as a method of control for mosquito larvae.

Gerris well deserves the popular name of pondskater, because it skims across the water at great speed. It moves in a series of glides, each following a propulsive stroke provided mainly by the powerful middle legs which sweep a wide arc on either side of the body. Only the feet of the front and middle legs, and the region below the knee joint (the tibia and tarsus) of the hind legs, are in contact with the water (fig. 18). These dimple the surface, an effect which is strikingly demonstrated by the shadows of a pondskater skating on clear shallow water over a pale background.

Between strokes, the front legs are held in front of the body with the feet resting on the water surface. At the beginning of the power stroke, they are lifted clear as the front of the body is raised slightly. The power stroke of the legs creates a pressure wave in the water surface (fig. 19, Nachtigall, 1974). The steep forward side of the wave is used like a runner's starting block against which the legs push the body forwards. The acceleration phase is followed by a glide phase which may last ten times as long as the active phase. High angular speeds are made possible by the small mass and the structural modifications of the leg. *Gerris* achieves maximum speeds of between 30 and 125 cm sec^{-1}, each push covering a distance of up to 15 cm. At the end of each stroke, the middle legs are lifted off the water and they return forwards through the air. The hind legs add little force to the push, but act as stabilisers (Nachtigall, 1974).

During a glide, the animal always moves in a straight line; it changes direction by moving the middle legs independently, as one uses the oars in a rowing boat. The insects orientate themselves rapidly in the direction of prey which is causing disturbance of the surface film. If startled, *Gerris* can jump as much as 10 cm into the air, using the middle and hind legs simultaneously, to escape from fish or other predators. On solutions of high viscosity, *Gerris* moves by walking or leaping, presumably because the power stroke is too weak to generate the surface wave in these conditions. Similar behaviour is observed in *G. lacustris* living in habitats polluted with discharges of thick oils. Adult gerrids may sink if the surface tension is lowered artificially with detergents to about 40–45 mN m^{-1}, but such low values are rarely recorded in the field (Brinkhurst, 1960). Stormy weather or ice cover drive *Gerris* to seek shelter on land, where their gait is slow and clumsy.

Gerris grooms itself frequently to remove debris which might clog the pile on its body, and to spread evenly the lipid secretions from a thoracic gland which may help to maintain the unwettable properties of the cuticle (Brinkhurst, 1960).

Many of the neustonic water bugs are less agile than *Gerris* and *Velia*. *Mesovelia*, *Hebrus* and *Hydrometra* are all to be found at the edge of reed beds on lakes and ponds and

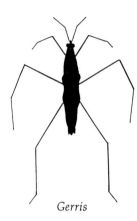

Gerris

Velia

amongst the emerging vegetation and floating leaves which give them more support than open water. All these have tarsal claws on the tips of the feet, which puncture the water surface to produce the forces required for locomotion. In *Gerris, Velia* and *Microvelia*, however, the claws are found just before the limb tip, which consists of a pad of water-repellent bristles in contact with the water surface (see p. 44). Macan (1976) relates this structural adaptation to the fact that they live on open water, leading a more truly aquatic life than their close relatives.

Early instars of pondskaters are generally found in dense vegetation near to the shore, whereas adults and larger nymphs occur on more open water. This habitat segregation of different developmental stages, which persists both seasonally and from year to year, is a world-wide phenomenon. Nummelin & Vepsäläinen (1988) found that the diet of gerrids varies unpredictably in time and space and does not explain the observed distribution of different developmental stages. Small instars prefer dense vegetation in which they are protected from predators (Spence, 1986) and can move freely, whereas the older long-legged stages need open water for easy movement.

Adult pondskaters and their relatives (Gerridae, Veliidae and Hydrometridae) are to be found in several forms: wingless, short winged or fully winged. The short-winged or wingless forms should not be confused with nymphs with wingbuds (see pl. 1). *Velia caprai* and the tiny *Microvelia* and *Hebrus* are all commonly wingless. Short-winged forms of *G. lacustris* occur frequently in the summer; their nymphal life is 5 days shorter than that of the fully-winged insects, which are commoner in the autumn. The advantages of a shorter generation time and faster population growth in the summer may be outweighed in the autumn by the benefit of being able to migrate on the wing. Winged forms may start new colonies in fresh habitats at any time of the year if the original habitat becomes overcrowded or polluted. Short-winged forms may be dispersed by water currents. The appearance of the different forms seems to be governed by environmental factors such as temperature and day length. The control is complex and may differ in different populations (Guthrie, 1959). Similar conclusions are reached for the American species *G. remigis* (Fairbairn, 1986).

Brinkhurst (1959a) studied *Velia* and *Gerris* in the field and found that in those with only one generation a year, wingless species occurred on open running water whilst fully-winged species were found on still-water habitats. In several species of *Gerris* having two generations a year, selection appeared to favour mixed summer populations of short-winged and fully-winged forms. The flightless forms may serve to maintain populations on existing suitable habitats; the fully-winged forms may migrate.

The adults overwinter as fully-winged forms capable of flight, but when the spring population is well-established, they resorb their wing-muscles, committing themselves to flightlessness, perhaps to increase the protein available for egg production.

Winged individuals generally fly only in the autumn. Flight is slow and they may be vulnerable to predators on land, although their scent glands make them unpalatable.

Erlandson and others (1988) observed *Velia caprai* in the field, and considered it to be a near-ideal species to study in the wild. They showed that *V. caprai* forages in groups, composed predominantly of females, with frequent interaction between individuals. Occasionally, short-lived territories were adopted and defended. The most frequent food source was adult flies, trapped in surface drift, and a few springtails were taken at night.

Further studies of the distribution and behaviour of pondskater populations in the field, and of the distribution in space and time of winged and wingless forms of various species, would be valuable.

4.5 Whirligig beetles

Gyrinus

The whirligig beetles, *Gyrinus*, give spectacular displays on the water surface. In late summer, they aggregate in schools which may contain hundreds of individuals, sometimes including more than one species. They are highly manoeuvrable and repeatedly change direction, following a characteristic and seemingly erratic path. Some species, such as *G. suffriani*, do not form schools, although there may be several individuals in the same area (Balfour-Browne, 1950). Whirligig beetles are widespread on still bodies of water and at the margins of slow-flowing rivers and streams, appearing to avoid vegetation or areas with visible scum. They are commonly found also on peat pools on high ground.

The body is oval and streamlined with a sharply-defined ridge right round the perimeter, clearly separating the body into upper and lower halves, and defining the level at which the beetle sits on the water. The eye is also divided into upper and lower halves, used for seeing in air and water respectively.

Because of the position of the body between air and water, the frictional drag due to contact with the water is greatly reduced. The beetle can swim about ten times faster on the surface than below it. The middle and hind legs provide propulsion while the forelegs are held forwards, ready to capture prey. With the hind legs beating 50–60 times a second and the middle ones at half this frequency, the beetle can achieve speeds up to $1m\ sec^{-1}$ – nearly 200 bodylengths a second – in short bursts on the surface (Evans, 1975).

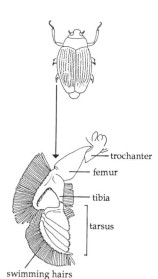

trochanter

femur

tibia

tarsus

swimming hairs

Fig. 20. The whirligig beetle
Gyrinus to show hind leg
adapted for swimming
(after Wigglesworth, 1964).

According to Nachtigall (1974), 'the leg of *Gyrinus* is the best propulsive mechanism working on the principle of resistance that has been found in the animal kingdom'. The flattened leg segments, fringed with swimming hairs, form effective paddles when the leg joints and swimming hairs are fully extended (fig. 20). The fringes allow feathering of the limb during the recovery stroke, when the blades close up like a fan and the tibia and tarsus fold into the femur. The whole limb is rotated to present a narrow edge with the minimum of resistance to the water (Evans, 1975).

When moving at high speed, the beetle creates up to 14 consecutive waves extending as far as six body-lengths in front of the insect. The short antennae project forwards, held exactly in the surface by their shape. Any object in the beetle's path will reflect the waves, which are detected by receptors in the antennae. *Gyrinus* can respond to this information so rapidly that it avoids solid objects with great accuracy even at night. It may avoid other beetles by responding to their waves rather than to the reflections of its own. A school of whirligigs gyrating at speed but apparently never colliding gives a remarkable demonstration of their navigational abilities.

An air store is carried under the wing covers when the beetle dives. When it stops swimming, its buoyancy ensures that it rapidly bobs up to the surface.

4.6 Spiders

There are a few species of spider which are semi-aquatic, always found near the water surface, as distinct from the fully aquatic 'water spider', *Argyroneta*. Bristowe (1958) suggests that *Argyroneta* shares with neustonic species the presence of a fluid produced by the spinnerets, which is spread over the legs and abdomen by the tarsi of the hind legs. The fluid is presumably water-repellent and keeps the spiders dry even when completely submerged. But *Argyroneta* is the only spider which can swim freely under water. Neustonic forms like *Dolomedes* and *Pirata* climb down plant stems to submerge and hang onto them under water. *D. fimbriatus*, variously called the 'swamp spider' or the 'European fishing spider', is a large spider and when it ventures onto the open water, the body, as well as much of the length of the legs, rests on the water. The diagonal walking pattern of the legs is replaced by a swimming pattern when the spider runs on water (Foelix, 1982). It rows across the water, with the legs extended and moved from the base, the two legs of a pair in synchrony. The second and third pairs of legs provide most of the power.

The female of 'nursery web' spiders like *Dolomedes* carries the large spherical egg-cocoon in her mouthparts (pl. 2.6). She spins a tent-like web in which the cocoon is deposited just before the young spiders hatch. They remain in the nursery web for several days before dispersing and beginning their life as predators. The smaller males are often eaten by the female, after mating.

Pirata piraticus and *P. piscatorius* both live in sphagnum bogs, beside pools of standing water. They sometimes build vertical silken tubes at the edge of the water with the upper end opening onto the water and the lower end leading down into the water (IV.4, p.60). The spider hides in the lower, cooler end, darting out to capture prey. According to Foelix (1982), they hunt by sight, reacting to prey movements. They are fierce hunters, like *Dolomedes*, and will catch tadpoles and small fish.

4.7 Flies

4.7.1 Adult flies

A sweep net passed through waterside or emergent vegetation will catch a great variety of adult flies. Many of these will be from nearby terrestrial habitats, but found by water occasionally when they have been blown there or strayed there by accident. A large number will be species with aquatic larvae whose adults are found on the surface only at the time of emergence from the pupae or when returning to lay their eggs. In particular, midges of the families Chironomidae, Chaoboridae and Ceratopogonidae are commonly encountered. Others whose larvae do not live in water are also seen on the surface, like the sciarid midges which may be abundant at times. Many flies are small and delicately built, with a large surface area so that they are vulnerable to desiccation. Near water, they may benefit from high humidity. A few groups have a closer, more specific relationship with the water surface. It is with those flies that the present account is concerned.

Most of the flies that frequent the water surface are predators. They feed mainly on small animals found at the surface, living or recently dead, such as larvae, waterfleas, newly-emerged midges and mosquitoes, and other Diptera. Particularly vulnerable are the insects that have fallen or been blown onto the surface, and newly-emerged adults before their skins harden.

There are many questions posed by predatory flies which could be answered by careful observation. Is there a regular sequence of different dominant predators throughout the seasons, reducing competition? How selective are they about prey? Is prey selection based on size alone, or on other factors such as seasonal availability? Do different predators have clearly defined, differing preferences?

Certainly, the flies appear to be at least as important in the predatory niches of the neuston as the more conspicuous beetles and bugs. The speed and agility of many flies, and especially their ability to become airborne in an instant, enable them to evade capture by large predators such as water boatmen, gerrids or spiders. Many of the predatory Diptera continue to hunt during late autumn and winter, when most of the larger predators have disappeared from the water and competition is reduced.

The Ephydridae (shore flies) are the commonest flies on the water surface, walking or hopping about boldly and found mostly on still water. Of the 130 species recorded in Britain, the most familiar species on still water belong to the genus *Hydrellia*, especially *H. griseola* and *H. modesta*. They are small greyish-green flies, 2–3 mm long. They catch their prey by tumbling it over and over with their legs until it becomes trapped in the water surface (Laurence, 1952). Hard teeth on the proboscis enable them to rasp soft-bodied prey such as other flies or springtails. As well as taking live prey, they are seen to probe the remains of insects left by other predators and also lower the proboscis into the film, perhaps to suck it up. Details of their feeding behaviour are incompletely known. *Hydrellia* is frequently found on temporary pools and puddles as well as permanent waters.

In the autumn, the large ephydrid *Ephydra micans* (length 5–6 mm) may be found on the surface. Although its sinister black appearance with slanting eyes and cowl-like head suggests a predator, it feeds by scraping off encrusting algae with ridges on its proboscis. *Ephydra riparia* is a coastal species, found on brackish pools.

Especially during the summer months, the long-legged Dolichopodidae are abundant at the water's edge, running on the surface or on wet mud. Most of them are larger than most ephydrids. They are typically 5–6 mm long and many have bronze or metallic green bodies. *Dolichopus* and *Poecilobothrus* species prefer to perch on emergent vegetation, alighting only momentarily on the water. The spectacular *P. nobilitatus* is seen in high summer, lying in wait on water lily leaves, or at the edges of cattle drinking places, to prey on waterfleas such as *Chydorus*, mosquito larvae or other flies (d'Assis Fonseca, 1978). Its black and white wing tips are effectively displayed during the semaphore-like movements of its wings during its courtship dance (Smith & Empson, 1955).

The smaller, darker dolichopodids *Hydrophorus* and *Campsicnemus* are less conspicuous. They can run on the water surface, where they abound. *Hydrophorus* species move very rapidly and are tricky to catch with a net. They hunt insects; *H. bipunctatus* can be found in late summer at the water's edge, feeding on aphids that fall from overhanging trees. *H. oceanus* is a coastal species. *Campsicnemus scambus* is a smaller, more delicate fly found particularly from September to December. It feeds on springtails and other insects. Both it and *Hydrophorus* are believed to suck up organic deposits from the film; much may still be discovered about their feeding behaviour.

The Empididae (dance flies) are generally found in terrestrial habitats but a few are associated with the water surface. *Hilara maura* forms conspicuous dense swarms between May and July, particularly over running water.

The flies pursue a triangular flight path a few centimetres above the water, occasionally snatching up insects from the surface by a sudden dart downwards, or plane across the surface, making a V-shaped wake. Small objects like grass flowers, tossed onto the surface, may divert a *Hilara* from the swarm. It seems to try to pick up a particle, perhaps mistaking it for prey. Males are observed to wrap prey in silk threads secreted by glands on the front feet, to offer to the female during the mating flight. A swarm of *Hilara* offers good hunting for larger predators. The larger empid, *Rhamphomyia sulcata*, particularly, may be caught holding a *Hilara*.

A number of empids seem to be scavengers of dead insects that have been trapped at the surface. *Clinocera stagnalis* is one of the species often found in summer and autumn, running on the surface at the margins of ponds and lakes. Laurence (1953) described it as 'the hyaena of the surface film'.

The rich source of prey on the surface film may sometimes attract predators from other habitats, perhaps when their usual supplies are short. The common yellow dungfly, *Scathophaga stercoraria*, is a good example. Generally found hunting on cow pats or flower heads such as hogweed, *Heracleum sphondylium*, it is occasionally seen hunting above the water surface.

Non-predatory regular visitors include the Sciaridae and Sphaeroceridae (cypselids or borborids in older reference books). These flies are seen to lower the proboscis into the surface, presumably to suck up nutritious decaying organic matter, together with the associated micro-organisms. Sphaerocerids such as *Leptocera humida* become prominent as the vegetation dies back, as very small blackish flies, gliding and hopping on the surface. Some species are noticeable in winter when there are few epineustonic animals about, although they occur in most other months also. Others are summer species. Those that glide do so in a manner suggestive of the use of a surfactant. The mechanism remains to be investigated. Another fly encountered occasionally at the surface film is the specialised muscid, *Lispe*. It has been collected in large numbers from shallow salt marsh pools.

As well as helping us to a better understanding of the biology of the neuston community, studies of predation of aquatic dipteran larvae and pupae, and the vulnerable emerging adults, are important for their possible relevance to the natural control of mosquitoes. Without the heavy predation which is known to occur, mosquito populations might be much larger.

4.7.2 Fly larvae

The larvae of Diptera may occur in enormous numbers even in small volumes of water such as rainwater tubs and garden ponds. Mosquitoes make good use of the surface at all stages of their life cycle.

Fig. 21. Egg raft of the mosquito *Culex*. Each egg has a flexible collar on the water surface (after Beament & Corbet, 1981).

Fig. 22. Larva of the mosquito *Anopheles*, suspended from the surface film, and feeding.

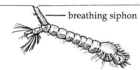

breathing siphon

Fig. 23. Larva of the mosquito *Culex*, hanging from the surface.

Culex species lay floating rafts of scores of eggs. Each egg is attached to the water surface by a flexible cone with a wettable lower surface and an unwettable upper one (Beament & Corbet, 1981). The shape adopted by the cone depends on the position of the egg in the raft (fig. 21). To watch a female adult *Culex* standing on the water and manipulating each egg into the raft as she lays it is to be a spectator at a wonderful display of juggling with surface forces.

Anopheles larvae feed on the 'ceiling' of the pond, as members of the hyponeuston. Their feeding brushes, fans of long hairs on the front of the head, sweep the underside of the surface film, directing a stream of particles towards the mouth (fig. 22). Useful food, including micro-organisms and particles of organic detritus, is filtered out and swallowed. *Anopheles* larvae are suspended parallel to the surface by several pairs of unwettable rosette-like hairs, held in position by surface tension (fig. 22). A pair of spiracles at the tip of the abdomen break the surface. An oily secretion repels water from their openings, which lead into the air-filled respiratory system. When the larva is supported at the surface, its head faces downwards. To feed, the larva rotates its head through 180° to bring the feeding brushes into position beneath the film.

Culex larvae feed deep in the water, surfacing from time to time to take in air through spiracles on the breathing siphon at the tip of the abdomen (fig. 23). This bears hinged hydrophobic lobes which are pulled apart by surface tension when the siphon meets the surface. The larva submerges when alarmed, detaching the siphon from the surface by a sudden flick of the feeding brushes. When the larva is immersed, the surface tension of the air–water interface of the bubble retained in the siphon now pulls the valves together and holds them shut.

Mosquito pupae are attached to the surface (fig. 8, p.7) by a pair of breathing trumpets on the thorax. Unlike most pupae, they are very active, swimming vigorously downwards by flapping the tail with its two broad swimming plates. The pupal stage lasts only a few days before the adult mosquito emerges. The pupa rises and as the cuticle over the thorax splits, the hydrophobic thorax of the emerging adult breaks the surface. The adult emerges into the air without getting wet. A thin film of surfactant secretion can be seen to spread from the pupal cuticle, pushing back particles on the surface.

Larvae of the highly-adapted meniscus midges (Dixidae) feed in a similar fashion to *Anopheles*. They take up a unique position with the body bent into a U-shape, held by the powerful surface forces in the meniscus in contact with leaves and stems at the water margin. The midge larva is held in position without effort as though by an elastic belt, while its mouth brushes work busily gathering in the film (fig. 24). Fine, long sensory bristles project from the tip of the abdomen and allow the midge

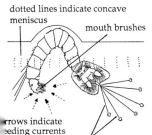

dotted lines indicate concave
meniscus

mouth brushes

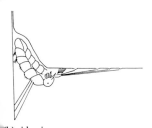

arrows indicate
feeding currents

sensory bristle, 'dimpling' the
water surface

(a) seen from above

(b) side view

Fig. 24. Larva of the meniscus
midge *Dixella* feeding in the
meniscus at the water's edge
(from original drawings by
D.M. Guthrie).

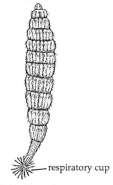

respiratory cup

Fig. 25. *Stratiomys* larva.

Fig. 26. Springing organ
(mucro) of the springtail
Sminthurides.

larva to detect the approach of a predator. If alarmed, the
larva swims off, flexing the trunk in a graceful beating
motion.

Stratiomys is another fly with a strikingly modified
aquatic larva. The tapering, flattened body ends in a circle of
branched, water-repellent hairs forming a respiratory cup
(fig. 25). Some beetle larvae, such as *Hydrobius*, may also be
found suspended at the surface in a similar fashion.

4.8 Springtails

Springtails aggregate in conspicuous swarms at the
margins of streams, ponds and lakes. If disturbed, they leap
suddenly in all directions and are much more difficult to
find. Their tiny size and water-repellent bodies are well
suited to neustonic life. The tips of the feet rest in
depressions on the surface, which is penetrated by the
hydrophilic tarsal claws (fig. 11, p.12). The blue-black
Podura aquatica and the creamy-white, humped
Sminthurides aquatica are anchored by the hydrophilic
ventral tube. They jump by means of a forked springing
organ (furcula) at the tip of the abdomen. When the insect is
at rest, the organ is folded forwards under the body and
held by a catch. When the catch is released, the springing
organ straightens out, slapping against the water and
projecting the animal through the air for a distance of about
15 body-lengths. *Sminthurides* increases the area with which
the springing organ strikes the water by a fringe of stiff hairs
with bent tips (fig. 26). Springtails exploit the fact that a
water surface does not deform rapidly, as anyone who has
played ducks and drakes with flat pebbles knows.

Podura aquatica has more than one generation a year.
Matthey (1971) noted that it was the first insect species to
appear on peat bogs in the spring, and the last to disappear.
The eggs were laid in the sphagnum, and a mixture of
adults and larvae were present in early summer, largely
disappearing in July. Populations reappeared in August.
Only adults were found in the winter. They remain active
under snow cover.

Another neustonic species of springtail, the blue-
black *Anurida maritima*, is easily recognised because it lacks
a springing organ and cannot jump. It has a widespread
distribution on brackish and saltwater pools.

Much remains to be discovered about the natural
history of neustonic springtails. Some feed on detritus and
on living components of the microneuston. Pollen is a major
food source at certain times of year when it sinks in clouds
onto habitats surrounded by trees such as willow and hazel.

Some questions that require investigation

How do neustonic animals ingest the surface film of microneuston? This could be investigated by seeing what happens to carmine particles or colloidal graphite ('Aquadag') scattered thinly on the surface. The film may be used by fine-particle feeders such as the rotifers *Brachionus* and *Keratella*, waterfleas and fly larvae, as well as by transient visitors such as snails. Remember that feeding habits may change with age: early larval stages may be particle feeders; later stages may be carnivorous.

Which micro-organisms are regular members of the microneuston? This could be approached by exploring the microneuston during long periods of calm weather. Observations on material in culture could be used to confirm the findings of field studies.

What do neustonic animals eat? This could be explored by looking at gut contents (in which it is possible to recognise the remains of algae and the hard parts of crustacea and insects); the remains of prey on the beaks of bugs; or the sucked-out skins or shells of prey floating on the surface in the wild or in cultures. Feeding habits of predators can be established by watching them carefully in the field and recording what they catch over a period of time. This can be followed up with experiments to discover which prey is preferred by predators when they are offered a choice in the laboratory. Feeding behaviour may differ from one habitat to another. Simple emergence traps may give information about the relative biomass of emerging insects available to predators. Some surface and aquatic bugs feed extensively on adult and larval mosquitoes; detailed knowledge of their feeding habits might be useful in the biological control of biting insects.

What is the seasonal sequence of predators on the surface in a particular habitat? The stability of the predator population in a habitat could be determined by making a census of flying immigrants and emigrants using a plastic or wooden quadrat frame floating on the suface.

How does a neustonic community develop over time? The colonisation of the water suface could be investigated by using light traps or white water traps (techniques, p. 68), or by creating a small new pond. Colonisation of temporary pools can also be studied. The neuston communities of shallow saltmarsh pools which are sometimes covered by the tide offer interesting possibilities.

Are pondskaters territorial? Marked specimens of *Gerris* could be used to discover the extent of an individual's territory. Does each insect have a regular 'beat'?

Is the occurrence of winged and wingless forms of water bugs seasonal, or related to the type of habitat in which they occur, or to the number of generations occurring in a year? *Gerris* has been much investigated (p. 23); what about some of the other common bugs?

How important is surface tension in supporting particular species? This might be explored by examining the effects of lowered surface tension on the ability of animals to stand on or hang from the surface. Contact angles of the body surfaces of different kinds of animal could be investigated (techniques, p. 77) to find out how the animal is supported at the surface. Experiments could be carried out on the effects of surfactants on the contact angles of insect cuticle, in relation to the insects' ability to move, and to receive information through their sense organs. We also need to know how far the presence of a particular species in different habitats is affected by decreases in surface tension, especially in relation to pollutants such as oil or surfactants.

Photography can be a valuable tool in the study of neustonic insects, which show up well against the water surface. The menisci formed where the limbs and body dimple the water can be observed by reflected light or as shadows on the bottom of an aquarium. A standard single lens reflex camera with extension rings can be used to obtain sufficient magnification, with or without flash. A home video can provide information about locomotor sequences, breathing activity and other behaviour.

How do neustonic communities on mildly polluted waters compare with those on clean waters? Mild pollution sometimes increases the species diversity of freshwater communities. By investigating the species composition of neustonic communities on many different habitats, and relating this to measurements of surface tension (techniques, p. 74) made in each habitat, it should be possible to discover how surface pollution affects the community, and perhaps to identify organisms in the micro- or macroneuston which can serve as pollution indicators. Any information about the effects of pollution is potentially useful.

'Jet propulsion' by secretion of surfactants (p. 15) may be commoner than we realise. Lycopodium powder or talc could be used to see which insects can move over the surface in this way.

What keeps 'schools' of whirligig beetles together? What determines their position on the pond? By marking individual beetles it may be possible to find out whether they school with the same companions repeatedly in distinct groups, or whether they mix freely throughout the whole habitat. Perhaps some environmental effect, such as a response to strong current, keeps groups together in regions of slack water.

Where do neustonic animals go in the winter? Which species overwinter as eggs, larvae, pupae or adults? It may be necessary to sample the population repeatedly as winter approaches to get clues about this. Where does hibernation take place?

5 Pollution

The organisms described in this handbook are generally found only in clean, unpolluted localities. The greatest diversity of species is usually found in the absence of pollution. It is well known that submerged aquatic animals are susceptible to toxic chemicals such as heavy metals, organic insecticides, fungicides and, to a lesser extent, herbicides. All of these may originate as industrial effluent or the run-off from adjacent cultivated land. The aquatic fauna may also be restricted by the shortage of oxygen that results from the decomposition of organic pollutants. No doubt members of the hyponeuston are similarly affected by soluble toxins and, to a lesser extent, by the shortage of oxygen that results from organic pollution. But a particular threat for neuston comes from surface-active pollutants (detergents, wetting agents from herbicide and pesticide formulations, surfactants applied in mosquito control) and non-polar oils.

In recent years, there has been much concern about the effects of oil on aquatic habitats. Most research has been devoted to large oil spillages at sea; less attention has been paid to effects on freshwater. Fuel and lubricating oils are now of particular concern on water bodies used for boating. Although modern lubricants contain surfactant additives, the main components of oil spillages on freshwater are non-polar long-chain hydrocarbons. Films of petroleum hydrocarbons and highly esterified oils do not spread easily because of their hydrophobic nature, combined with strong intermolecular attraction. They may form slicks of hundreds of molecules in thickness, showing the characteristic rainbow interference colours. The main cause of damage to neustonic animals is mechanical, due to the high viscosity and adherent properties of the oil. It will stick to the hydrophobic cuticle of insects, clogging the hair piles necessary for locomotion and respiration, and blocking the spiracles through which the insect breathes. It will also interfere with emergence and egg-laying. Jahn (1972) investigated the effects of petrol on waters used by pleasure craft in Germany. He found that thin layers of oil, even in scattered patches, destroyed a considerable part of the fauna immediately. All water bugs, some beetles and their larvae, and many fly larvae, succumbed. Meniscus midge larvae (*Dixa* species) and some dytiscid beetles proved more resistant.

One hopeful development is the production of a biodegradable, nontoxic lubricant (Castrol Biolube 100) for use in outboard motors. This provides an alternative to heavy lubricating oils, especially welcome in areas where intensive leisure use of the water poses conservation problems.

A secondary effect of oil is to exclude oxygen from the water, affecting all those animals which depend on dissolved oxygen for their needs. The fauna may also be damaged by the water-soluble toxic fractions of petrol. Jahn (1972) found that animals not in direct contact with the oil film were endangered by toxic ingredients at concentrations above 0.5 ml/1 water. Similar effects on fourth instar larvae of the mosquito *Aedes aegypti* were investigated by Berry & Brammer (1977) and Berry, Brammer & Bee (1978). Toxicity of different fractions increased as water solubility increased with the series benzene, toluene, xylene. Larvae were especially vulnerable when young or during moulting, which suggests that the cuticle is a barrier against oils entering the body. Toxic effects of heavy bunker oil on river plankton are also reported by McCauley (1966).

Small spillages of oil probably have only transient effects. Hydrocarbons will gradually be degraded by bacteria and by ultraviolet and visible light, or dispersed more rapidly on open water by wind and wave action. Baier (1972) considers that aerosol formation due to bubble-breaking is the most efficient process for dispersing hydrocarbon films. Jahn (1972) observed that snails and mussels actually fed on the oil, which helped to disperse it. They did not appear to suffer ill-effects as they were later able to reproduce. Marine protozoa will ingest crude-oil residues together with normal food sources both in the field and in culture (Andrews & Floodgate, 1974).

Oil spillages may be dispersed by emulsifiers which are detergents developed by industry specifically for the purpose of removing oil. These are typical man-made surfactants. Their hydrocarbon chains stick to the hydrocarbon chains in the oil, and their polar groups are attracted to the water (fig. 9, p.8). The oil is broken up into droplets which can be suspended in the water as an emulsion and washed away. But man-made (or indeed, natural) surfactants pose other problems for the surface fauna. Insects which inhabit the surface, emerge through it, or lay eggs on it may be at risk from anything which lowers the surface tension, or which may adsorb onto the body surface, changing its properties. As well as attaching themselves to oil droplets, the hydrocarbon chains of detergent molecules may attach themselves to the hydrocarbon chains of the waxes which cover the cuticle of many insects. Surfactant molecules accumulate at the air–water interface and will creep onto the hydrophobic surfaces of insects and stick there, with their hydrophilic (polar) ends outwards. These hydrophilic groups are now orientated on the outside of the insect surface, lowering the contact angle to less than 90°. In other words, the previously hydrophobic areas of the insect's body are now wettable, with unfortunate results for all those functions which depend on a high contact angle for water. The surface forces on which many insects depend for support will be greatly altered, which may make it difficult for them to stand, walk or skate on the surface. Also, much of the information received by the insect from its environment is detected by sense organs which depend on the displacement of bristles by disturbance of the meniscus. Predators depend on such information for detection of prey (see p. 20). Many epineustonic insects have sensory receptors in the legs, monitoring their position in relation to the surface. Whirligig beetles avoid collisions by detecting ripples with sense organs in their antennae (see p. 26). Changes in surface tension and contact angles will affect the deflection of sensory bristles and may provide incorrect information from any of these receptors. Very few such systems have been investigated in detail. Studies are needed to find out more precisely how changes in surface tension and wettability affect the life of the neuston. Such investigations could provide models to assist in monitoring pollution of this kind.

The foaming rivers of the early 1960s were caused by the accumulation of synthetic detergents from industrial and domestic effluent. Since those early detergents have been replaced by kinds with much lower levels of foaming agents, foaming is no longer a common problem, except downstream of points of effluent discharge. In addition, the general use of biodegradable detergents, which can be broken down by micro-organisms in the environment, has eliminated the damaging surfactant effects of detergents which persisted in the environment. Eutrophication due to phosphate derived from detergents remains a problem.

In arid regions, long-chain hydrophobic alcohols such as hexadecanol and octadecanol are spread on reservoirs as evaporation supressants. As surfactants, they form monomolecular layers on water and reduce evaporation by 30–50%. They are supposedly biodegradable, but affect the surface communities by increasing bacterial growth and raising the water temperature just below the surface. They also affect algal populations and shore vegetation.

The surface effects that contribute so much to pollution have long been harnessed in the control of malaria-carrying mosquitoes. In the 1930s, tropical swamps were coated with a thick layer of mineral oil to kill the immature stages. After the second world war, dramatic control was effected by applying insecticides such as DDT from the air. Such methods are used less now, partly because many mosquitoes have developed insecticide resistance and partly because of undesirable effects on other organisms. Modern methods of control include application to freshwater habitats of surfactants, such as the commercial surfactant known as Monoxci. It is non-toxic and spreads as a monolayer lasting only a few days. It appears to work by coating the hydrophobic respiratory opening of the pupa, making it more wettable. The pupa is prevented from making contact with air at the water surface (McMullen, Reiter & Philips, 1977). These methods perhaps cause less environmental damage, but their effects on components of the neuston other than mosquitoes remain to be explored.

Most insecticides are highly soluble in oil but not in water. The same is true of the toxic polychlorinated biphenyls (PCBs) widely used in industry as solvents and in the manufacture of plastics. Such lipophilic toxic substances, as well as heavy metals (lead, mercury, copper, chromium and zinc), are avidly collected into petrol films. Heavy metals are often present naturally at low concentrations in the water, and may also occur as pollutants. They are adsorbed, together with other contaminants, onto natural surfactants such as proteins. These collect at the gas–water interface of bubbles forming in the water, which, when they burst at the surface, will concentrate such pollutants there. If these enter the neuston food chain, they may be further concentrated at higher trophic levels.

Much remains to be discovered about the action of pollutants which accumulate naturally at the water surface and the response of the neuston community to them. A review of literature and some tentative conclusions are to be found in Jones and others (1980), but more research is needed. Laboratory studies of effects on particular species could be related to field observations and measurements of surface tension. Because the neuston is so readily visible, it could be particularly valuable as a biological index of water pollution if enough knowledge were available.

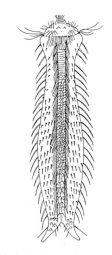

I.1

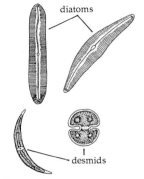

I.2 *Chaetonotus maximus*

diatoms

desmids

I.3

6 Identification

The neuston includes representatives of many different groups of organisms. This chapter will help to identify them to a taxonomic level that varies from group to group. For instance, relatively few species of water bug are neustonic, and many of these can be named to species using Key IIB backed up, where necessary, by the more detailed key in Savage (1989). On the other hand, many species of protozoa and rotifers are occasionally found in the neuston, and reliable identification of these to species is not possible in a book of this length. Key V and the Guide to some rotifers (p. 64) are therefore not comprehensive keys to genera, but frameworks for the description of some examples illustrated in this chapter and pls. 4, 6 and 7. Critical identification of the species in these groups would require help from an expert. Table 1 shows how some of the major groups of neustonic animals are classified. Beginners should start with Key I.

The animal should be set up so that its structure can be seen clearly. This means mounting it on a microscope slide under a compound microscope, or viewing it at lower magnification by transmitted or reflected light under a stereoscopic (dissection) microscope, or using a lens or the naked eye. Some groups, such as waterfleas, include forms of a range of sizes. These groups should key out whether or not your specimen is small enough to require the use of a compound microscope. Streble & Krauter (1973) illustrate many microscopic freshwater organisms and that book is very useful for approximate identification, even for those who do not read German. Fitter & Manuel (1986), Macan (1959) and Croft (1986) are useful for larger organisms.

I Key to animals found at the water surface

1   Animals barely visible to the naked eye; needing a compound microscope for clear observation. (View these under a compound microscope to proceed further with this key.)   2

–   Animals visible to the naked eye, but generally needing a lens or dissecting microscope for critical observation 15

2   Bearing beating cilia (fine hair-like processes) or knobbed tentacles (I.1)   3

–   Neither beating cilia nor tentacles visible   6

3   With a wheel-like ring of long cilia at one end (you may need to wait a while for the ring to be extended)   4

–   Without a ring of cilia at one end   5

4   Body covered with a delicate, colourless cuticle which
    gives the body a definite shape, and with a pair of jaws,
    the mastax, deep in the body; the mastax is easy to see if
    the jaws are making regular grinding movements, but
    may be inconspicuous if not   rotifers, see Guide, p. 64
–   Body naked, without even a delicate cuticle
                                        peritrich ciliates, Key V

I.4 A rotifer, contracted

5   Forked at the rear end (I.2)   Gastrotricha
–   Not forked at the rear end   ciliate Protista, Key V

6   Single-celled   7
–   Many-celled   8

7   Coloured cells (green, brown or yellow) with a clear cell
    wall, or a box-like structure which maintains a
    permanent shape; no structures for locomotion present(I.3)
        microscopic plants (diatoms, desmids and other 'algae'
        including cyanobacteria) (see also couplet 13)
            Belcher & Swale (1976) illustrate and name the
            common genera of freshwater algae.
–   Colourless, or if coloured, with a few whip-like flagella
    or lobe-like pseudopodia, with which the organism can
    move   other Protista, Key V

pseudopodia: cytoplasmic
extensions which appear to
flow out of the body

8   Body enclosed within a flexible or jointed cuticle; no
    two-valved shell   9
–   Body enclosed within a shell or case which is simple or
    two-valved   13

9   Body not obviously segmented   10
–   Body segmented   12

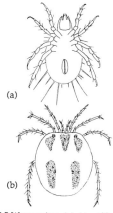

(a)

(b)

I.5 Water mites. (a) *Atractides*
nymph; (b) *Hygrobates* adult

10 Without legs   contracted rotifer (I.4) (see Guide, p.64)
–   With jointed legs   11

11 Two eyes; six (in juveniles) or eight (in adults)
    unbranched legs (I.5)
                        juvenile or adult water mites, Key IV
–   One eye in the middle of the head; six or eight limbs,
    some of them generally branched (I.6)
                        nauplius larvae of copepod crustaceans

I.6 Nauplius larva of a copepod

12 Never more than three pairs of jointed limbs (body
    usually divided into three clear regions [head, thorax
    and abdomen])   insects, Key II
–   More than three pairs of jointed limbs (body not divided
    into head, thorax and abdomen)
                        copepods and their larvae, Key III

I.7 Resting eggs of a rotifer,
Brachionus quadratus
(length 0.14 mm)

I.8 *Sphaerium corneum*

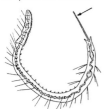

I.9 *Stylaria lacustris*
(length 15 mm)

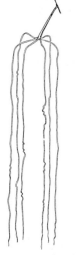

I.10 A hydra, *Chlorohydra
viridissima*, fishing

I.11 *Pisidium amnicum*

13   Shell with two valves hinged together   14
–   Shell or case single, without an opening
                rotifer egg (I.7), mosquito egg (pl. 8.5) or
                polyzoan statoblast (pl. 8.1, 8.2) (see also testate
                amoebae, Key V, couplet 5)

14   Without jointed limbs; one or two delicate tubes
     (siphons) and a lobe-like foot may emerge from the shell
                        immature bivalve molluscs (I.8)
–   With jointed, bristly limbs attached outside the shell or
     sticking out from it when the animal is moving
                        microscopic crustaceans, Key III

15   Tadpole! (pl. 8.4)   a tadpole
–   Not like this   16

16   Body soft, deformable, naked   17
–   Body with a shell or cuticle (which may be delicate and
     flexible)   19

17   Body segmented, with a row of bristles (chaetae) along
     each side   oligochaete worms
                (*Stylaria* (I.9) can be recognised by the threadlike
                filament at the front end)
–   Body not segmented and without chaetae   18

18   A flattened, creeping animal with two or many eyes
     (dark spots) at the front (pl. 8.7)
                        a flatworm (Platyhelminthes)
–   Not flattened; a simple, sac-shaped animal with up to 12
     tentacles at one end; green or brown (I.10)   a hydra

19   With a coiled spiral shell   a snail (Mollusca) (pl. 8.3)
–   Without a coiled spiral shell   20

20   Shell of two valves, hinged together   21
–   Without a two-valved shell   22

21   With a single fleshy foot emerging from the shell
                        a bivalve (Mollusca) (I.11)
–   With many jointed limbs, visible through the semi-
     transparent shell or with the tips sticking out; or attached
     to the head outside the shell   crustaceans, Key III

22   With jointed legs   23
–   Without legs or with stumpy unjointed legs (prolegs)
                        insect larvae or pupae, Key II

23   With eight legs   arachnids, Key IV
–   With six legs   insects, Key II

II Key to insects

Adult, larval or pupal insects found on the water surface, or flying just above it. Useful general books are Chinery (1976, 1986) and Merritt & Cummins (1978).

Nymphs and larvae

In this book, the term 'larvae' refers to the immature feeding stages of endopterygote insects (members of orders like the flies and beetles, in which the developing wings are not visible from outside and a pupal stage is present). 'Nymphs' are the immature stages of exopterygote insects (orders like the bugs and mayflies with external wing buds and no pupal stage). Some authorities refer to the immature feeding stages of all insects as larvae.

1	Wings or wingpads present; forewings may be leathery or hard and shell-llike, concealing hindwings
–	Wings or wingpads entirely absent

1	Wings or wingpads present; forewings may be leathery or hard and shell-llike, concealing hindwings	2
–	Wings or wingpads entirely absent	7

2	With fully developed wings which can be moved	3
–	With short wingpads which cannot be moved	5

3	Forewings not leathery or hard	4
–	Forewings leathery or hard, at least in the basal half (nearest the body)	6

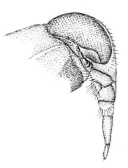

II.1 Side view of thorax of fly (front end to left)

4	Hindwings reduced to small, knob-like halteres (II.1) adult flies (Diptera), Key IID
–	Hindwings present other insects

These include caddisflies (order Trichoptera), mothlike insects with the wings clothed in tiny hairs and folded rooflike over the back. A few caddisflies walk on the water surface, particularly members of the family Hydroptilidae.

II.2 Heteropteran rostrum: *Notonecta*

5	With legs; mouthparts forming a beak (II.2) short-winged or nymphal water bugs (Heteroptera) Key IIB
–	No legs; without beaklike mouthparts pupae or puparia of flies (Diptera), Key IIF

6	Forewings forming hard wingcases (elytra) which meet in the midline along the back and cover the hindwings which are folded beneath them; conspicuous jaws used for biting beetles (Coleoptera), Key IIC

(In camphor beetles the elytra are much shorter than the abdomen.)

– Forewings leathery in basal half (or more); sucking mouthparts forming a beak

water bugs (Heteroptera), Key IIB

II.3 (after Chinery, 1976)

7   With a ventral tube (II.3) projecting below the first
    abdominal segment; abdomen with six or fewer
    segments; (rarely more than 3 mm long)
                          springtails (Collembola), Key IIA
–   No ventral tube; abdominal segments, if visible, more
    than six; (insect usually more than 6 mm long)   8

8   Without free, jointed legs   9
–   With three pairs of free, jointed legs   11

9   With jointed legs stuck down against the body; usually
    brownish   pupae of flies (Diptera), Key IIF
–   Without jointed legs, although fleshy unjointed
    projections called prolegs may be present   10

10 Soft, flexible and usually pale
                          larvae of flies (Diptera), Key IIE
–   Hard, rigid, amber or dark brown
                          puparia of flies (Diptera), Key IIF

11 Middle and hind legs long and slender, extending
    considerably beyond abdomen; compound eyes present;
    mouthparts forming a beak
                          wingless water bugs (Heteroptera), Key IIB
–   Legs not longer than the abdomen; compound eyes
    absent   larvae of beetles (Coleoptera)
    (These are larvae such as *Dytiscus*, *Hydrophilus* and
    Hydrobius, which come to the surface to breathe, but
    which spend most of their time in deeper water and
    are therefore not included in the key.)

IIA   Key to springtails (order Collembola)

Springtails are commonly found in large numbers at the
margins of ponds, ditches and slow-flowing rivers. They are
among the most constant members of the neustonic
community, although unless they are active they may go
unnoticed because they are so small. There are many
terrestrial species; Fjellberg's (1980) key covers many British
species. They often land on the water surface. The following
key works only for those that are found in large groups and
clearly at home on the water surface; it does not cater for
terrestrial forms that have arrived on the surface by
accident.

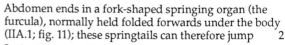

IIA.1 (after Chinery, 1976)

1   Abdomen ends in a fork-shaped springing organ (the
    furcula), normally held folded forwards under the body
    (IIA.1; fig. 11); these springtails can therefore jump   2
–   Springing organ absent; (springtails found in the
    intertidal zone, in estuaries and rockpools)
                          Anurida maritima (Laboulbène)

IIA.2

2 Body rounded, almost spherical; springing organ ends in flattened, paddle-like lobes (IIA.2)
Sminthurides aquaticus (Bourlet) (pl. 5.1)
– Body elongate; springing organ without flat lobes
Podura aquatica L. (pl. 5.2)

IIB Key to water bugs (order Heteroptera)

This key deals with adult bugs. All bugs with full-length wings are adults. Bugs with short wings may be adults of a short-winged form, or they may be nymphs. Nymphs of aquatic and semiaquatic bugs can be distinguished from short-winged adults because nymphs have only one tarsal segment on each foot, whereas adults have at least some legs with two tarsal segments; and because nymphs are pale on the underside whereas adults are dark or have at least some dark patches there. This key will take the commonest bugs of the water surface to species; others can be identified further with the excellent keys and illustrations in Savage (1989), Macan (1976) and Southwood & Leston (1959). For keys to nymphs, see Brinkhurst (1959b) and Mitis (1937).

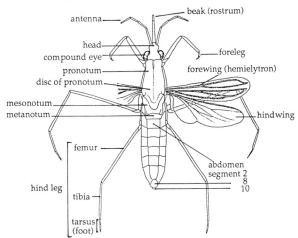

IIB.0 Glossary diagram: structure of a water bug (*Gerris*) (after Macan, 1976)

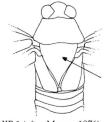

IIB.1 (after Macan, 1976)

1 Antennae shorter than head; living *in* the water 2
– Antennae conspicuous, longer than the head; living *on* the water 3

2 Mouthparts consisting of a sharp, pointed beak; scutellum (a triangular plate between the wings and the pronotum, IIB.1) visible, not concealed by wings; length 13–16 mm
water boatman *Notonecta* species (family Notonectidae)

Notonecta swims powerfully, upside down, using oar-like hind legs fringed with swimming hairs. Of the four British species, the commonest is *Notonecta*

glauca L., which is abundant and widespread
throughout Britain. It is the only species with the
forewings entirely pale except for the black markings
confined to the front margin (pl. 2.4).

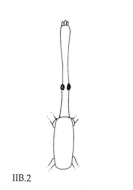

–   Mouthparts a blunt, stubby beak; scutellum usually
concealed by wings

               lesser water boatman (family Corixidae)

There are many species of corixid in Britain. They
spend most of their time feeding on the bottom
deposits, but come to the surface occasionally to
breathe air.

3   Head many times longer than broad (IIB.2), with eyes
placed some distance from the thorax; body very thin
and slender, narrowing to elongate head; walks slowly
on the water

IIB.2

     water measurer *Hydrometra* (family Hydrometridae)

There are two British species. *Hydrometra stagnorum*
(L.) (pl. 2.7) occurs throughout Britain amongst reeds
at the edge of slow-flowing waters. Its body is 9–12
mm long and blackish-brown. The rarer *H. gracilenta*
Horvath is recorded only from the Norfolk Broads
and the New Forest. Its body is 7.5–9 mm long and
red-brown or yellow-brown.

 –   Head as broad as long or nearly so; eyes close to front
border of thorax   4

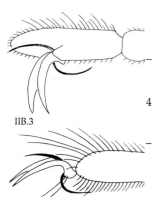

4   Antennae of four segments of about equal thickness;
claws inserted before the tip of the foot, at least on the
IIB.3   foreleg (IIB.3) (use a lens magnifying at least x 10)   5

 –   Antennae of five segments, the last three thinner than
the first two; claws of all legs inserted at tips of the feet
(IIB.4)   *Hebrus* (family Hebridae)

IIB.4

There are two species of *Hebrus* in Britain. The
common   *H. ruficeps* (Thomson) (pl. 1.2) (length
1.2–1.5 mm) is usually wingless and is found in
sphagnum at the margins of acid ponds and tarns.
The rarer *H. pusillus* (Fallén) (length 1.5–2 mm),
recorded only from Wales and the southern half of
England, is distinguished from *H. ruficeps* by a white
spot and a white streak on the minute forewings.

5   Hind femur extending well past the tip of the abdomen

          Pondskaters or water striders (family Gerridae)   6

 –   Hind femur short and not extending beyond tip of
abdomen   14

6   Disc (a large swelling in the middle of the rear of the
pronotum) (IIB.5) and first three segments of antennae
yellowish or reddish brown   7

 –   Disc of pronotum uniformly dark   9

PLATE 1

Hemiptera (water bugs)

1. *Microvelia reticulata* (x10)

2. *Hebrus ruficeps* (x10)

3. *Gerris lacustris*
   nymph (x3)

4. *Gerris lacustris*
   fully-winged form (x3)

5. *Gerris lacustris*
   short-winged form (x3)

6. *Gerris argentatus* (x3)

7. *Gerris costai* (x3)

8. *Velia caprai*
   wingless form (x3)

9. *Velia caprai*
   winged form (x3)

10. *Gerris najas* (x3)

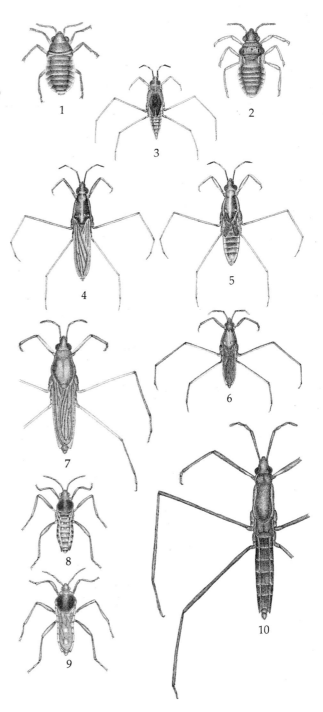

PLATE 2

Water bugs, water beetles and spiders

1. *Gyrinus caspius*
   whirligig beetle (x7)

2. *Gyrinus marinus*
   whirligig beetle (x7)

3. *Gyrinus substriatus*
   whirligig beetle (x7)

4. *Notonecta glauca*
   water boatman (x2.5)

5. *Pirata piraticus*
   spider (x2.5)

6. *Dolomedes fimbriatus*
   spider (x1.5)

7. *Hydrometra stagnorum*
   water measurer (x2.5)

8. *Dianous coerulescens*
   camphor beetle (x5)

9. *Helophorus aequalis*
   water beetle (x7)

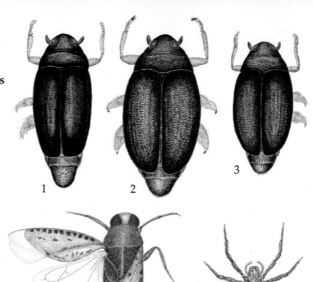

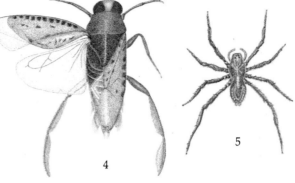

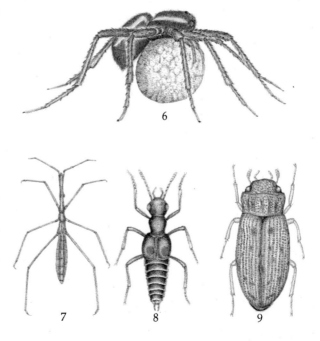

PLATE 3

Diptera (flies)

1. *Hydrellia griseola* (x10)

2. *Poecilobothrus nobilitatus* (X5)

3. *Leptocera (Opacifrons) humida* (x10)

4. *Clinocera stagnalis* (x6)

5. *Hilara maura* (x6)

6. *Campsicnemus scambus* (x5)

7. *Lispe tentaculata* (x5)

8. *Hydrophorus bipunctatus* (x5)

9. *Ephydra riparia* (x6)

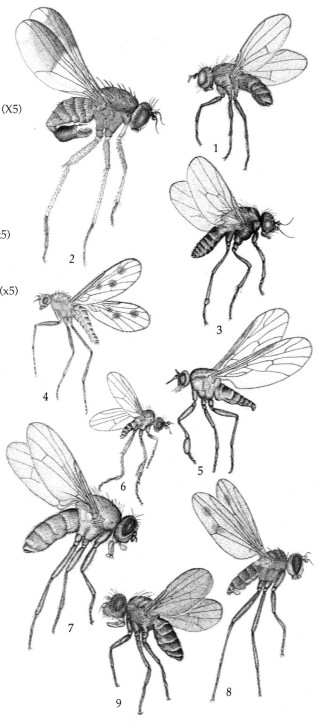

PLATE 4

Protozoa

1. *Euglena*
   (20–500 μm)

2. *Arcella*
   (30–260 μm)

3. *Anthophysa*
   (individuals 5–6 μm)

4. *Synura*
   (colony 100–400 μm)

5. *Trachelomonas*
   (20–45 μm)

6. *Phacus*
   (30–150 μm)

7. *Chilomonas*
   (30–50 μm)

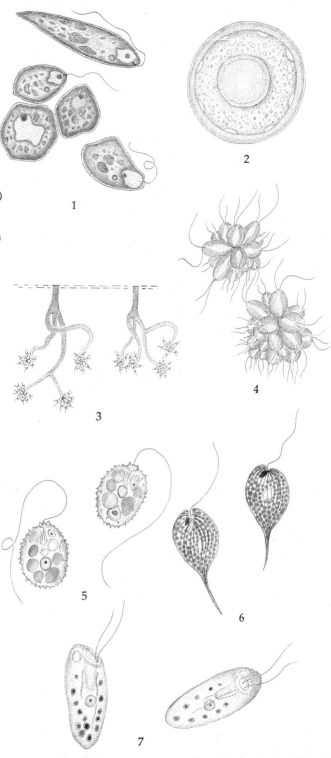

PLATE 5

1. *Sminthurides aquaticus*
   springtail

2. *Podura aquatica*
   springtail

3. *Anopheles claviger*
   anopheline mosquito
   pupa

4. *Anopheles claviger*
   larva

5. *Aedes punctor*
   culicine mosquito
   larva

6. *Aedes punctor*
   pupa

7. *Dixella aestivalis*
   meniscus midge
   larva

8. *Dixella aestivalis*
   adult

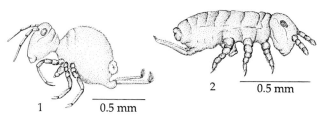

1   0.5 mm

2   0.5 mm

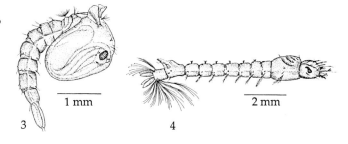

3   1 mm

4   2 mm

5   2 mm

6   1 mm

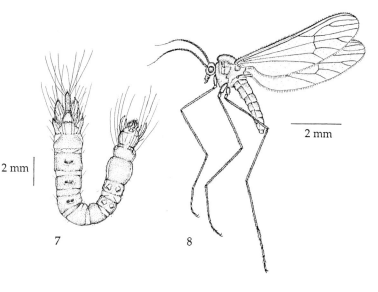

2 mm

7   8   2 mm

PLATE 6

Microneuston

1. *Notodromas monacha*
   ostracod
   female

2. *Notodromas monacha*
   male

3. *Scapholeberis mucronata*
   waterflea

Rotifers

4. *Brachionus*

5. *Filinia*

6. *Collotheca pelagica*

7. *Euchlanys dilatata*

8. *Rotaria neptunia*

9. *Squatinella rostrum*

10. *Keratella cochlearis*

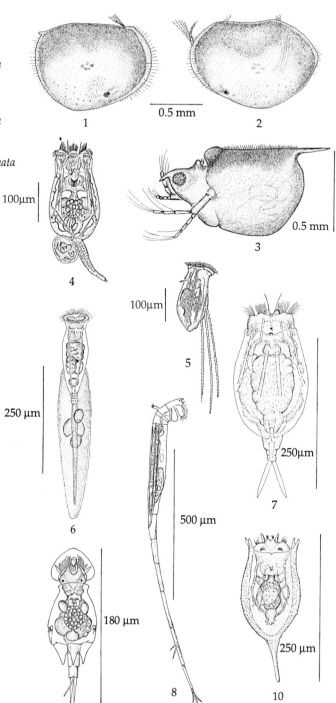

PLATE 7

Protozoa

1. *Stylonychia*

2. *Sphaerophrya*

3. *Podophrya*

4. *Euplotes*

5. *Oxytricha*

6. *Chilodonella*

7. *Vorticella*

8. *Aspidisca*

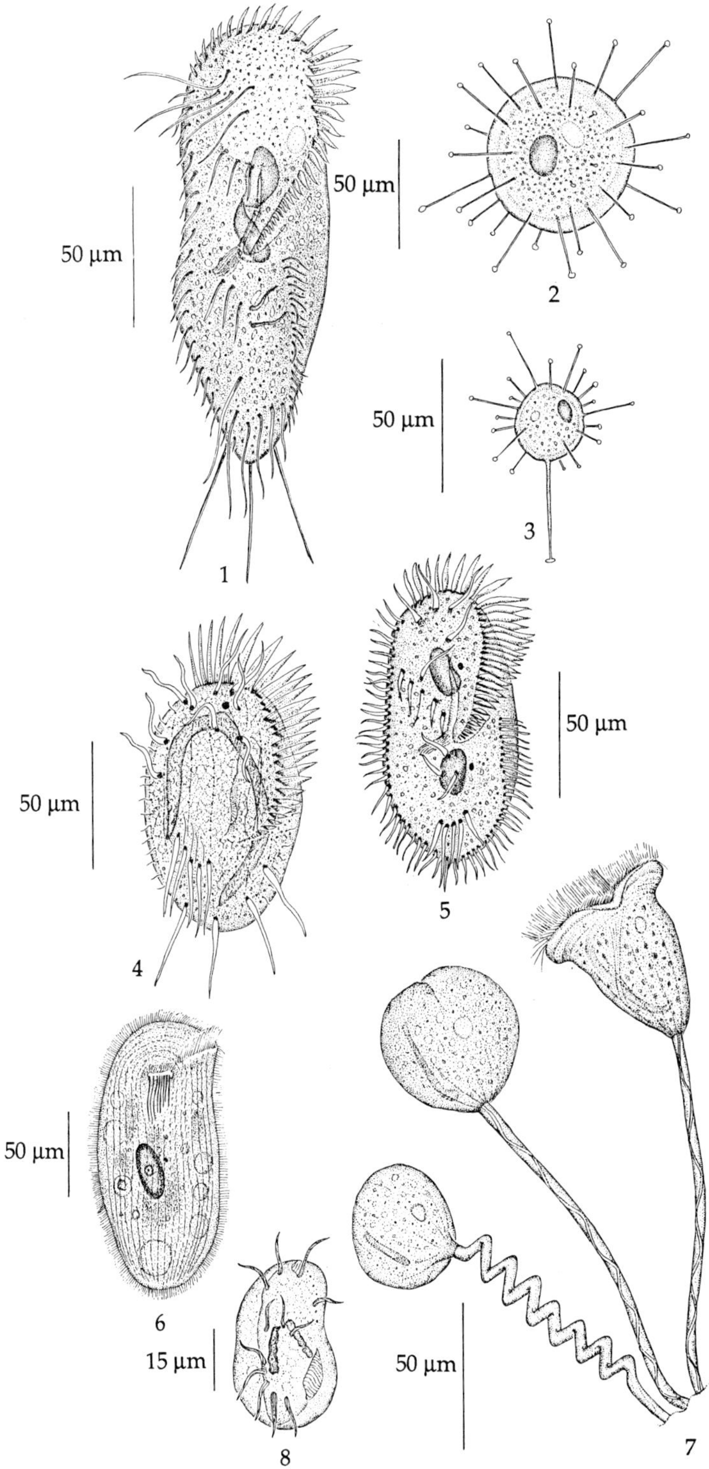

PLATE 8

1. *Cristatella mucedo*
    statoblast of polyzoa

2. *Plumatella*
    statoblast of polyzoa

3. *Lymnaea*
    snail

4. *Rana temporaria*
    tadpole

5. *Culex* species
    mosquito egg raft

6. *Tintinnidium fluviatile*

7. *Polycelis* species
    flatworm

8. *Tetanocera ferruginea*
    puparium

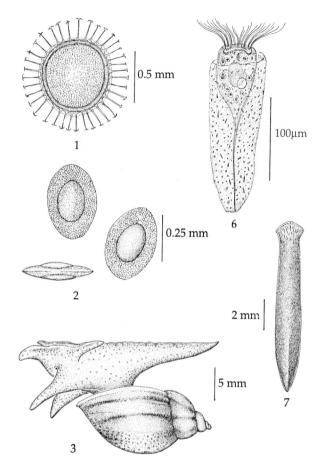

0.5 mm

1

100µm

0.25 mm

6

2

2 mm

5 mm

7

3

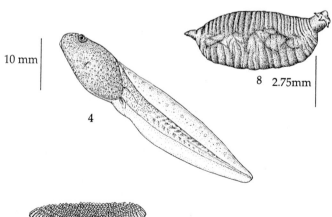

10 mm

8   2.75mm

4

5

7 Body broad across disc of pronotum and bases of middle legs; small, dark brown patch (often hard to see) at front of pronotum behind eye; length 12–14 mm
Gerris costai (Herrich-Schaeffer)* (pl. 1.7)
Usually winged. Common, especially in the north; moorlands and upland tarns, sometimes in brackish habitats.

– Body more or less parallel-sided; no small brown patch behind eye   8
If specimen has abdominal spine (IIB.6, IIB.7), see note in couplet 10

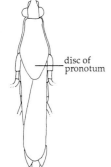

disc of pronotum

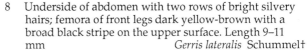

8 Underside of abdomen with two rows of bright silvery hairs; femora of front legs dark yellow-brown with a broad black stripe on the upper surface. Length 9–11 mm   *Gerris lateralis* Schummel†
(previously known as *G. asper* (Fieber). A local species.)

– Underside of abdomen without two rows of bright silvery hairs; length 10–12 mm
Gerris thoracicus Schummel (IIB.5)
Widely distributed and common, tolerating brackish conditions and organic pollution.

IIB.5 Outline of *Gerris thoracicus* (after Macan, 1976)

9 First antennal segment longer than the second and third together; sixth visible abdominal segment in both sexes with spines at the hind corners (IIB.6)   10

– First antennal segment as long as, or a little shorter than, the second and third together; sixth abdominal segment in both sexes angular, but not extended into a spine at the sides (IIB.7)   11

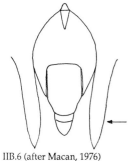

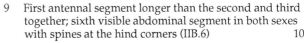

10 Spines at hind corners of sixth visible abdominal segment not reaching tip of abdomen (IIB.7); large, 13–17 mm   *Gerris najas* (De Geer) (pl. 1.10)
Usually wingless. Widely distributed on larger bodies of water; ponds, lakes and rivers; prefers flowing water.

IIB.6 (after Macan, 1976)

– Spines at hind corners of sixth visible abdominal segment reaching as far as, or beyond, the tip of the abdomen; yellow line on sides of pronotum; usually winged; length 14–16 mm   *Gerris paludum* (Fabricius)
South-east England only, on bodies of still water.

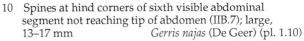

Another large gerrid with abdominal spines is the rare *Limnoporus rufoscutellatus* (Latreille), distinguished by its red-brown pronotal disc.

IIB.7 (after Macan, 1976)

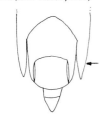

11 Metasternum with a tiny yellow bump; yellow line on the side of the pronotum not continued in front of constriction (IIB.8); length 10–13 mm
Gerris gibbifer Schummel
A broad, rather dark species, widespread, common and abundant on acid waters including peat pools.

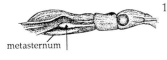

metasternum

IIB.8 (after Macan, 1976)

* *G. costai poissoni* Wagner & Zimmerman in Savage (1989)
† *G. lateralis asper* (Fieber) in Savage (1989)

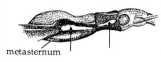

metasternum

IIB.9 (after Macan, 1976)

– Metasternum without yellow bump; yellow line on the side of the pronotum continued in front of constriction (IIB.9)   12

12 A line of silvery hairs around the hind edge of the pronotum; (hind tibia and tarsus together about 2/3 as long as the femur); length 6.5–8 mm
              Gerris argentatus Schummel (pl. 1.6)
      Usually winged. Distribution in Britain restricted to England and Ireland, especially amongst reeds, sometimes in brackish habitats.

– No silvery hairs on the pronotum   13

IIB.10 (after Macan, 1976)

13 Femur of foreleg pale, with a black band along the front running from tip about two thirds towards base, and another black band on the back running from tip to middle; length 8–10 mm   *Gerris lacustris* (L.) (pl. 1.5)
      The most widespread and abundant, and the most pollution-tolerant, of the British species; in ponds and ditches, lakes and backwaters. Wingless, short-winged and long-winged forms occur.

– Femur of foreleg black except for the basal half of the undersurface; (males with a blunt tooth on either side of the sixth visible abdominal segment, IIB.10); length 7–9 mm   *Gerris odontogaster* (Zetterstedt)
      A dark species; widespread, common and abundant on weedy still waters, often with *G. lacustris* although it appears to be less tolerant of pollution.

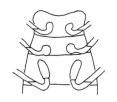

IIB.11 Underside of thorax, *Mesovelia* (after Macan, 1976)

14 These insects are generally wingless. Legs attached near the midline of the body (IIB.11); length 3–3.5 mm   *Mesovelia furcata* Mulsant and Rey
      The only British species of the family Mesoveliidae. It is greenish with black markings on the upper surface. Not truly a neustonic species, being generally found on the leaves of water plants like *Potamogeton*.

– Legs attached at sides of body (IIB.12)
                                          Family Veliidae 15

IIB.12 Underside of thorax, *Velia* (after Macan, 1976)

15 Length 6–8 mm   adult water cricket *Velia* species
      There are two British species. *Velia caprai* Tamanini (pl. 1.8, 1.9) (length 5.5–7 mm) is very common and abundant throughout the British Isles on small bodies of clean, still or gently flowing water. Its pale or reddish pronotum distinguishes it from the less common *V. saulii* Tamanini which has a black or dark brown pronotum. Colour is an unreliable character; Macan (1976) gives other features which separate males and wingless females of the two species. Winged females of the two species cannot be distinguished on external features. *V. saulii* (length 6–6.8 mm) prefers larger bodies of water.

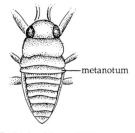

metanotum

IIB.13 (after Macan, 1976)

– Length 1.4–2.5 mm   16

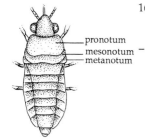

IIB.14 (after Macan, 1976)

16   Body uniformly black; metanotum (covering rear
     thoracic segment) clearly visible (IIB.13)
                                    Velia nymphs (see above)
–   Patches of white hairs on thorax and abdomen;
     metanotum visible only at the corners, the rest being
     covered by the pronotum and mesonotum (IIB.14)
                                    Microvelia species

     Microvelia reticulata (Burmeister) (pl. 1.1) (length 1.4
     mm) is common throughout Britain, in sheltered
     habitats amongst reeds and sedges at the edges of
     lakes, pools and ditches. It tolerates slight salinity.
     Two other species, of very limited distribution, can be
     identified using Macan (1976) or Savage (1989).

IIC Key to adult water beetles (order Coleoptera)

        The order Coleoptera, one of the largest orders of insects
in Britain, includes a large number of freshwater species,
many of which are occasionally caught at the water surface.
Many water beetles visit the surface periodically to
replenish their air supplies. Also found at the surface are
numerous species of tiny beetles which are not aquatic at all.
        Since this book cannot deal with all of the many aquatic
species, it deals instead with three selected groups adapted
to life at the water surface and regularly found there. A
specimen that matches the description of one of these
groups in every respect can be identified further here. Other
specimens caught at the surface can be identified in the
AIDGAP key to British water beetles (Friday, 1988) or, if
they are not truly aquatic forms, in Unwin (1984). Many are
illustrated in Harde (1984).

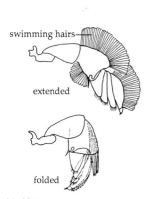

hind leg

Whirligig beetles (family Gyrinidae)

        These are usually recognised by their characteristic
behaviour when at the surface, where they skate rapidly
around with the body half submerged, changing direction
abruptly, and skilfully avoiding obstacles and other beetles.
They are generally found in 'schools' of one or two to many
hundreds of individuals, often of more than one species.
Gyrinids are characterised by the following combination of
features:

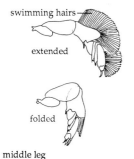

middle leg

IIC.1 (after Nachtigall, 1961)

        middle and hind legs shorter than forelegs, broad and
        flat, able to fold up like a fan (IIC.1);

        each compound eye divided horizontally into two
        distinct parts (IIC.2).

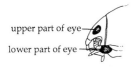

IIC.2

        The 12 British species resemble one another closely and
are hard to separate except by examining the genitalia,
which are illustrated by Friday (1988). This is a shortened
version of Friday's key, the full version of which should be

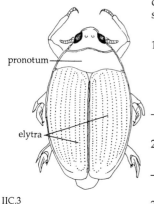

pronotum

elytra

IIC.3

consulted for further identification. Among the commonest species in Britain are *Gyrinus substriatus* and *G. marinus*.

1   Pronotum and elytra (IIC.3) covered in fine, short hairs (best seen when the beetle is dry)
                              Orectochilus villosus (Müller)
    A nocturnal species found at the surface only at night and remaining submerged during the day.
–   Pronotum and elytra without hairs   2

2   Underside entirely yellow-red
              Gyrinus minutus Fabricius or *G. urinator* Illiger
–   Underside mainly dark, with some parts orange   3

3   Claws of middle and hind feet black (use a lens), much darker than the rest of the leg
              Gyrinus marinus Gyllenhal or *G. aeratus* Stephens
    G. marinus (pl. 2.2) is widely distributed, especially in brackish and peaty waters.
–   Claws of the middle and hind feet yellow, the same colour as the leg   4

4   Pronotum and elytra dull (when beetle is dry)
                              Gyrinus opacus Sahlberg
–   Pronotum and elytra shiny (when beetle is dry)   5

5   Elongate beetles with sides of the elytra almost parallel in the middle; elytra scarcely wider at their mid-point than at their front margin (pl. 2.1)
              Gyrinus caspius Menetries (pl. 2.1) or *G. paykulli* Ochs
                    (formerly called *G. bicolor* Fabricius)
    G. caspius is common near the coast.
–   Rounded beetles; elytra at widest point about 1.25 times wider than at the front margin (pl. 2.3)
              Gyrinus suffriani Scriba, *G. distinctus* Aube, *G. natator*
                    (Linnaeus) or *G. substriatus* Stephens (pl. 2.3)
    G. substriatus is generally common and abundant.

Helophorus (family Hydrophilidae, subfamily Helophorinae)

IIC.4

Small beetles (1.9–7.0 mm) characterised by the *five wide, wavy grooves or furrows* (pl. 2.9) that run the full length of the pronotum (IIC.4) and are clearly visible with a lens. The palps are often longer than the antennae, and are easily mistaken for them. The palps arise from the mouthparts and have three segments; the antennae arise from the head itself and have more segments. In *Helophorus* the antennae are clubbed and they and the palps can be tucked into a groove on the head.

The 17 species of aquatic *Helophorus* are keyed out in
Friday (1988). One of the commonest in Britain is
H. brevipalpis Bedel, in which the last segment of the palps is
symmetrically spindle-shaped, whichever angle you view it
from, and much longer than broad, *and* in which the rows of
pits near the midline all run almost the full length of the
elytra. *Helophorus aequalis* Thomson is another of the
commoner species and is illustrated in pl. 2.9.

Camphor beetles (family Staphylinidae, subfamily Steninae)

Beetles of the family Staphylinidae are easily recognised
by the very short elytra, which leave several abdominal
segments exposed, antennae with more than five segments,
and (four or) five segments on each foot. There are nearly
1000 British species in this family, and they are difficult to
separate. Beginners are advised not to tackle this family
without help.

The subfamily Steninae contains two genera, *Stenus*
with 72 British species and *Dianous* with only one species,
D. coerulescens (Gyllenhal) (pl. 2.8). This species has a
distinctive blue colour with a yellow spot on each of the
elytra. It lives in the moss and under stones near waterfalls
and emerges to hunt springtails on the water; rather rarer
than the other beetles described above.

IID Key to families of adult flies (Diptera) of neustonic habits

Almost any fly may be found having fallen or blown
onto the water surface. Unsuccessful struggles to escape
probably indicate species not 'at home' in this habitat. Many
flies have immature stages which are aquatic, and the adults
may be seen or captured on the water as they emerge and
fly (such as the enormous numbers of Chironomidae, non-
biting midges, and Culicidae, mosquitoes). This key is
confined to a small number of forms belonging to a few
families where careful observation shows the adults to have
a clear and regular association with the water surface –
walking, hopping or gliding over it, or flying continuously
just above it. It does not include the many flies which may
visit the surface on a facultative or irregular basis.

The Diptera is a large and varied order, many members
of which are difficult for the non-dipterist to identify to
species, and specialist help may therefore be necessary for
critical identifications. Unwin (1981) provides a full key to
Dipteran families, as well as much useful practical advice.
Colyer & Hammond (1968) is a mine of information and has
a useful appendix on collecting, preservation and
examination. The Royal Entomological Society Handbooks
and the Freshwater Biological Association Scientific
Publications provide keys to some families.

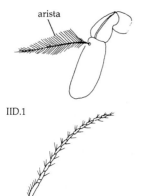

IID.1

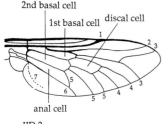

IID.2

2nd basal cell
1st basal cell discal cell

anal cell

IID.3

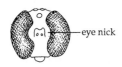
eye nick

IID.4

1st basal cell

cross vein

IID.5

1st basal cell
2nd basal cell

anal cell

IID.6

1   Antennae with fewer than five segments, relatively short and stubby, or with a whip-like structure (the arista) arising from the third (last) segment of the antenna proper (IID.1)   2
–   Antenna threadlike, of at least five segments (IID.2) (Suborder Nematocera, including gnats, mosquitoes and craneflies)   16

2   Anal cell long and pointed (IID.3), often reaching the wing margin; both basal cells long also
                                                  non-neustonic families
–   Anal cell relatively short, extending less than 2/3 the distance to the wing margin, or absent altogether; second basal cell usually short   3

3   Cross veins crowded into basal quarter of wing, the tip of which may be pointed   non-neustonic families
–   At least one cross vein beyond basal quarter of wing, the tip of which is always rounded   4

4   Arista, when present, at tip of third antennal segment   5
–   Arista situated on upper face of third antennal segment before tip   7

5   Inner margin of eye with a nick at level of base of antenna (IID.4)
                        empids Empididae (part) (see next couplet)
–   No such eye nicks   6

6   Only one cross vein in outer three quarters of wing (IID.5). Usually metallic, shining (bluish, greenish or bronze) flies Dolichopodidae (part only) (see couplet 8)
–   At least two cross veins in outer three quarters of wing (IID.6). Flies never shining metallic
                              dance flies Empididae (part)
    Some members of the Empididae are found in huge numbers at the water surface. The large genus *Hilara* includes numerous species which skim just above the surface, weaving back and forth at great speed.

Examples

Hilara maura Fallén (pl. 3.5). Forms conspicuous mating swarms in June, of many hundreds of flies following an elliptical horizontal flight path. It is 3.5–4.5 mm long; males with first tarsal segment of front leg much dilated (about 3 x width of tibia); a spot (stigma) near leading edge of wing; dagger-like mouthparts; thorax and halteres blackish, the former with three indistinct stripes, and appearing greyish at sides; knees yellow. On the wing May–July.

Clinocera stagnalis Haliday (pl. 3.4). 4 mm long; body grey-lilac (like watered silk); wings indistinctly clouded; proboscis very short; third antennal segment short and with downwardly-directed arista at tip; elongated front coxae; tips of tarsi with three pads; long bristle at base of wing (humeral). No long bristle below base of front femur). On the wing May–September, along with other members of the genus. Collin (1961) and Chvála (1975, 1983) key the species.

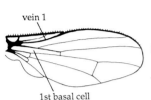

vein 1

1st basal cell

IID.7

7   First basal cell (IID. 5) very short (at most half as long as vein 1) or absent   8

–   First basal cell well over half length of vein 1 (IID. 7)   9

8   Bristles immediately behind each eye on back of head strongly developed (IID. 8)

dolichopodids or 'dollies' Dolichopodidae (part)

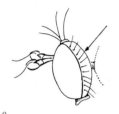

IID.8

Examples

Poecilobothrus nobilitatus L. (pl. 3.2). 5.6–6.0 mm long; green metallic body with striking black and white wing tips; arista with long hairs. June–July on waterweed and surface of pools and ditches.

Hydrophorus species. For example *H. bipunctatus* Lehmann (pl. 3.8). 4.0–5.0 mm long; grey-brown body; wing with two dark spots; male front femur with 8–14 spines beneath. On water March–October.

Campsicnemus specie. For example *C. scambus* Fallén. (pl. 3.6). 2.5–3.25 mm long; males with contorted legs, the middle tibia being dilated, curved and spiny and the mid-tarsus short, deformed and ornamented; face ochre-yellow, frons (above face) shiny violet blue; arista inserted near base of third antennal segment. Runs on surface of pools and ditches throughout the year. d'Assis Fonseca (1978) keys the Dolichopodidae.

–   These bristles minute, hair-like, or absent   9

9   Hind tibia with strong bristles as well as those at tip or near tip   10

–   Hind tibia with strong bristles near the tip only   14

10 Back of head with fine pale hairs below level of neck (and frequently elsewhere as well)

large dung flies Scathophagidae

The common yellow dung fly *Scathophaga stercoraria* L. is an occasional predator at the water surface. *Cordilura* species also occur where there are sedge swamps. Collin (1958) keys the Scathophagidae.

–   Back of head without such fine pale hairs   11

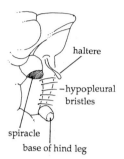

haltere

—hypopleural bristles

spiracle

base of hind leg

IID.9 (after Chinery, 1976)

11   A row of bristles between the base of hind leg and spiracle situated just below the base of haltere (IID. 9)
non-neustonic families
—   This area (the hypopleuron) devoid of bristles   12

12   The next-to-last longitudinal vein reaches wing margin, even though outer half may be faint
non-neustonic families
—   Neither of the last two longitudinal veins reaches wing margin   13

13   The last two longitudinal veins are divergent or almost parallel in outer halves
houseflies and relatives Muscidae
The subfamily Lispinae are characterised by the expanded spoon-shaped palps on their mouthparts. *Lispe* is a common predatory genus of the neuston, including several marine littoral species.
L. tentaculata Lehmann (pl. 3.7) has distinctive grey chevrons on the abdominal tergites.
d'Assis Fonseca (1968) keys the Muscidae.
—   These two veins are convergent in outer halves
non-neustonic families

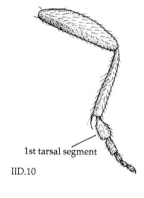

1st tarsal segment

IID.10

14   The first tarsal segment of hind leg at most 1.5 times length of second segment and usually fatter (IID.10). The antennal arista is often especially long and fine (being 1–1.5 x breadth of head)
Sphaeroceridae (=Borboridae or Cypselidae)
Small dark, typically spiny, flies. Some species always found near water. For example *Leptocera (Opacifrons) humida* Haliday (pl. 3.3) – a blackish fly with strong veins; vein 2 extends well beyond cross vein in outer half of wing; vein 3 almost straight; and tarsal segment of front leg with a single long hair near middle. Found on the water surface, being especially noticeable in winter. Pitkin (1988) keys the species.
—   The first tarsal segment of hind leg clearly more than 1.5 times length of second segment and not fatter   15

breaks in front margin

hindmost cross vein

IID.11

15   The hindmost cross vein is at or beyond middle of wing. Vein forming front margin of wing usually with two obvious breaks in basal half (IID.11). There is no small bristle on top face of hind tibia just before tip. There are bristles on the sides of thorax above the base of middle leg. There are no bristles behind the ocelli (simple eyes on top of head in middle). When long hairs are present on arista of antenna they are restricted to the upper side; head relatively large with large mouth and olarge upper lip   ephydrids or shore flies, Ephydridae

Examples

Hydrellia species. There are two dozen British species, primarily distinguished by the male genitalia. Two widespread species are noted below.

Hydrellia griseola Fallén (pl. 3.1). 1.8–2.5 mm long; front of head below antennae gold; area just above antennae silver-grey; third antennal segment black; silvery-grey body, darker above. Very abundant throughout most of the year, walking and hopping on the surface.

Hydrellia modesta Loew. 1.8–2.5 mm long; very similar to above in appearance and habits, but front of head below antennae is silver.

Ephydra riparia Fallén (pl. 3.9). 4.5 mm long; greenish metallic reflections on body; head large, especially in male, with a very large mouth giving a cowl-like appearance to the front of the head; arista with fine hairs on the upper surface; without a long hair close to the arista; wings brownish. Often present in large numbers on brackish pools.

Consult Colyer & Hammond (1968) for species identification.

- Without this combination other families (of Acalyptrata)

16 With nine or more veins reaching the wing margin (not counting the small cross vein near the wing base)
non-neustonic families
- Fewer than nine veins reaching the wing margin   17

17 Antennae inserted below level of compound eyes
non-neustonic families
- Antennae inserted between compound eyes   18

18 Tibiae with strong spurs at tips (IID.12)   19
- Tibiae without spurs or they are clearly shorter than diameter of tibia   20

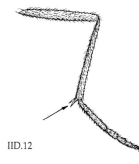

IID.12

19 Eyes meeting to form an 'eyebridge' above the antennae (IID.13)   lesser fungus gnats Sciaridae

Species of *Bradysia, Sciara* and other genera are 3–7 mm long; generally blackish; with a somewhat humped look to the thorax above the head. They are to be observed, on occasions, feeding on the organic deposits in the microlayer. Freeman (1983) keys the species
- Eyes well separated   non-neustonic families

IID.13

20 Eyes meet above the antennae to form an 'eyebridge' (IID.13)
non-neustonic families

- Eyes well separated, even when the beginnings of an 'eyebridge' are evident   21

21   With bottle-brush antennae (with whorls of long hairs
     on each segment) and a pair of claspers at tip of
     abdomen   male Chironomidae (see next couplet)
  –   Without bottle-brush antennae (at most with only one
     whorl of long hairs on each segment). Abdomen with or
     without claspers at tip   22

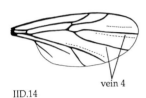

22   Vein 4 usually obviously forked (IID.14). Proboscis long
     (at least 3/4 length of palps)
                       females of biting midges Ceratopogonidae
  –   Vein 4 never forked (in some, venation is greatly
     reduced). Proboscis short (at most only half length of
     palps)   non-biting midges Chironomidae
     This is a large family (about 500 species in about 120
     genera in Britain). The majority have aquatic larvae.
     Emerging adults usually pause only briefly at the
     surface, but large numbers fall prey to both fish and
     neustonic predators. Some members of the subfamily
     Orthocladiinae appear to be neustonic, spending
     much time walking across the surface.
     Pinder (1978) keys the males to species.

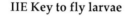
vein 4

IID.14

IIE Key to fly larvae

The larvae of midges and mosquitoes may occur in
enormous numbers and make up a major element of the
hyponeuston. A key to the principal families and higher
groups is given by Croft (1986). A fuller treatment is given
by Smith (in press). Keys to the species level are available in
the Freshwater Biological Association Scientific Publication
series for Culicidae, Dixidae and Chironomidae (in part).
Rozkošný (1973) deals with Stratiomyidae, Brindle (1960,
1967) with Tipulidae and Cranston and others (1987) with
mosquitoes. For other families see Smith (in press) or
consult a specialist for advice.

1   Larva with extensible tail which is always more than
    half the length of rest of body   2
  –   Larva usually without such a tail, or with tail less than
    half the length of rest of body. (When, rarely, it is about
    half the length then there is no distinct head capsule –
    see couplet 5)   3

2   Head reduced and indistinct, whitish like body, which
    abruptly narrows to tail
    subfamily Eristalinae of the family Syrphidae (hoverflies)
  –   Head a distinct dark brown to black structure. Body
    greyish and only gradually narrowing to tail (IIE.1)
                       Ptychopteridae (black craneflies)

IIE.1

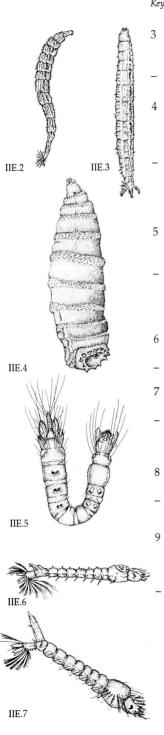

IIE.2

IIE.3

IIE.4

IIE.5

IIE.6

IIE.7

3 Larva tapered at both ends and with a tuft of hairs at the tail end, by which it will hang suspended from the surface film (IIE.2)   Stratiomyidae (soldier flies)

– Larva not of this form   4

4 Larva with a dark head which can be withdrawn into the front segment of the thorax, and tail end with breathing pores (spiracles) surrounded by lobes (IIE.3)

Tipulidae (craneflies)

– Larva not of this form   5

NOTE. If the tail has such lobes but the head is not darkened, apart from the protruding tips of the hook-shaped mandibles, go to 5.

5 Head in the form of a distinct hard capsule, which cannot be withdrawn into the front thoracic segment. The head is black, brown or orange to yellow   6

– Head reduced, and not in the form of a capsule. Only the tips of the protruding, hook-shaped, blackish mandibles are hardened (IIE.4)

Cyclorrhapha (higher flies)

6 Unjointed legs (prolegs), with short spines or hairs at tips, present on thorax and/or abdomen   7

– No prolegs present   8

7 A pair of prolegs present on each of first two abdominal segments (IIE.5)   Dixidae (meniscus midges)

– A pair of prolegs present on both the thorax and the last abdominal segment

Chironomidae (non-biting midges)

8 Abdominal segments clearly longer than wide. Larva usually yellowish   Ceratopogonidae (biting midges)

– Abdominal segments not longer than wide. Larva usually brownish, greyish or blackish   9

9 Abdominal segments each with three short, but wide, blackish plates on top, each of which has one or more rows of strongish hairs. Thorax not wider than abdomen

Psychodidae (moth flies or owl midges)

– Abdominal segments without three hairy plates on top, Thorax clearly wider than abdomen (IIE.6, IIE.7)

Culicidae (mosquitoes)

Members of the subfamily Anophelinae hang with the body parallel to the water surface suspended by rosettes of branched hairs on the abdomen, and by the spiracles on the eighth abdominal segment (IIE.6). Members of the subfamily Culicinae hang suspended at an angle to the surface by the respiratory siphon situated at the tail end (IIE.7).

IIF Key to fly pupae

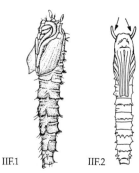

IIF.1   IIF.2

IIF.3

IIF.4

IIF.5 (after Redfern, 1975)

IIF.6 (after Marshall, 1938)

IIF.7 (after Miall, 1934)

1 With sheaths of developing adult legs stuck down against the body, and also with wing pads extending from thorax to beginning of abdomen (IIF. 1)   2
– No leg sheaths or wing pads   11

2 With a pair of breathing horns on thorax (one is sometimes much longer than the other) (IIF.2, IIF.3)   3
– Each breathing horn is replaced by a tuft of whitish filaments (IIF.4)   some Chironomidae

3 Leg sheaths straight (IIF.2, IIF.5)   4
– Leg sheaths with tips curving round hind margins of wing pads   8

4 Ends of leg sheaths extend well beyond hind margins of wing pads   5
– Ends of leg sheaths extend only about as far as hind margins of wing pads   7

5 One breathing horn much longer than the other (IIF.3)   6
– Both breathing horns about the same length (IIF.5)   some Tipulidae

6 The longer breathing horn is longer than body (IIF.3)   Ptychopteridae
– The longer breathing horn is less than three quarters length of body   some Tipulidae

7 Tips of wing pads slightly, but distinctly, pointed   Psychodidae
– Tips of wing pads evenly rounded   Ceratopogonidae

8 A pair of paddles at tail end, each of which has a distinct midrib (IIF.6)   9
– When paddles are present at tail end they lack a midrib   10

9 Breathing horns widest in middle and then tapering to pointed tips (IIF.7)   Chaoboridae
– Breathing horns widest at tips (IIF.8)   Culicidae
   (For further identification see Cranston and others, 1987)

10 Breathing horns expanding to tips, which form open cups (IIF.9)   Dixidae
– Breathing horns narrowing towards closed tips   some Chironomidae

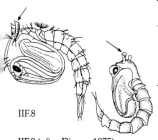

IIF.8

IIF.9 (after Disney, 1975)

IIF.10

IIF.11

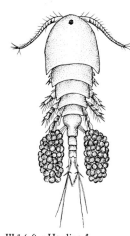

IIF.12

III.1 (after Harding & Smith, 1974)

11　More elongate puparium with a surface made up of minute hexagonal plates each bearing a central conical point, so that the texture of the surface is very rough
Stratiomyidae

－　Nearly always more barrel shaped, and so broader in relation to the length (for example IIF.10). Surface of puparium generally smooth　　Cyclorrhapha

Most will need rearing through to the adult in order to take the identification further. Some have the rear end bent upwards, like the prow of a viking ship, so that the breathing pores (spiracles) are held clear of the water surface. If such puparia are orange brown to almost black they are probably Sciomyzidae (IIF.11). If pale yellowish and somewhat translucent they are probably Scathophagidae. Puparia with the hind spiracles at the end of the branches of a backward-pointing forked tail (IIF.12) are probably Ephydridae.

III Key to microscopic Crustacea

All but the largest microscopic crustaceans should be mounted on a slide in a drop of water under a coverslip and viewed with a compound microscope. Permanent preparations can be made by mounting the specimens in polyvinyl lactophenol, which takes a few hours to clear them. Larger opaque forms like *Notodromas* are best seen by reflected light using a stereoscopic (dissecting) microscope.

1　Body and legs enclosed within a two-valved carapace (shell)　　2

－　Body and legs not enclosed in a two-valved carapace; the legs and the segmentation of the body can be seen; body pear-shaped, tapering and ending in a forked tail; single eye; antennae extend sideways and are used as 'balancers' or for slow swimming; females often carry one to two prominent egg sacs (III.1)
copepods (subclass Copepoda)

Copepods are not true members of the neuston, but are so numerous in the plankton that they are likely to occur in samples from the surface as temporary inhabitants. Their identification is a challenging task involving delicate dissection of the limbs (Harding & Smith, 1974). The immature stages (nauplii and copepodites) cannot be identified to species. *Tropocyclops prasinus* Fischer is a copepod which lives in small ponds and swims on its back. It hangs from the surface film by its first antennae but does not feed there (Dr G. Fryer, unpublished information).

2   Two-valved carapace bean-shaped, enclosing head as well as body, and usually concealing antennae and legs when at rest; antennae and first pair of legs emerge and move the animal in a jerky fashion; average size about 1 mm   ostracods (class Ostracoda)

   The only true neustonic ostracod in Britain is *Notodromas monacha* (O. F. Müller) (pl. 6.1, 6.2). The shell valves have a flattened ventral margin, armed with rows of simple hairs, by which the animal suspends itself from the meniscus. The colour pattern is characteristic. A patchy dark coloration due to minute pigment granules packed in large numbers just beneath the shell is more strikingly developed on the ventral surface (uppermost when the animal is suspended from the surface). *Notodromas* has well developed eyes, whereas most other ostracods have a single pigment spot. Groups of *Notodromas* move just below the surface in the same way as schools of whirligig beetles move just above it.

–   Two-valved carapace does not enclose the head or the antennae, which are conspicuous
                       waterfleas (order Cladocera)   3

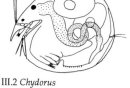

III.2 *Chydorus*

   Cladocera are a dominant component of the plankton in most bodies of freshwater, and will frequently be collected as temporary inhabitants of the neuston. They can be identified in Scourfield & Harding (1966). The tiny *Chydorus sphaericus* O. F. Müller (III.2) is common and widely distributed, but not truly neustonic. Dr G. Fryer reports that *Chydorus* and the larger *Eurycercus* are often found trapped in the water surface by surface forces; daphniids and bosminids are also vulnerable. The only waterfleas that are truly neustonic are members of the subfamily Scapholeberinae (family Daphniidae), in which the genera *Scapholeberis* and *Megafenestra* are adapted for life in the hyponeuston. They have straight ventral margins to the carapace, by which the animals can hang, back downwards, below the surface. There is a band of dark pigment along the ventral edge of the carapace in *Scapholeberis*; its possible function is discussed on p. 20.

III.3

3   Ventral margins of carapace valves straight, ending in a sharp angle or spine (III.3)   subfamily Scapholeberinae   4
–   Not like this   other Cladocera

4   Rostrum long and pointed (III.4); a window-like plate on the dorsal surface of the head shield (use high magnification and critical illumination); body about 1.5 mm long   female of *Megafenestra aurita* (Fischer)
III.4   *M. aurita* is much less common than *S. mucronata*.

III.5

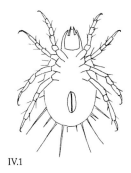

IV.1

IV.2 (after Lockett &
Millidge, 1951)

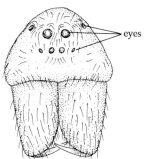

IV.3 (after Lockett &
Millidge, 1951)

– Rostrum short and blunt (III.5)

> *Scapholeberis mucronata* O. F. Müller (pl. 6.3) or
> males of *M. aurita*

Nearly all individuals that key out here will be
females of the common and widely distributed
Scapholeberis mucronata (about 1 mm long). These
should match pl. 6.3 exactly except for the length of
the spine on the head and that on the carapace. The
spines show seasonal variation in length (Green,
1963). Specimens not matching the plate will be the
rarely-found males of *S. mucronata* (up to 1 mm) or
M. aurita (about 1 mm), which can be separated from
one another under a compound microscope with
critical illumination by the presence in *M. aurita* of an
oval window-like plate on the dorsal surface of the
head shield (Dumont & Pensaert, 1983).

IV Key to spiders and mites (class Arachnida)

1   Body globular, round or oval, often brightly coloured,
    orange, red, green, blue or brown (usually about
    1–5 mm long)   water mites (order Acariformes)

Water mites are not normally collected with neuston,
but as active swimmers they may sometimes be found
at the surface. The larvae, with three pairs of legs, are
ectoparasites of water bugs and other insects, and
may be found in considerable numbers attached to
pondskaters. The larva metamorphoses into a
freeliving, carnivorous nymph, occasionally found at
the surface after leaving the larval host. The nymph of
Atractides (IV.1) is found among floating waterweed;
it has long hairs which may act as flotation devices at
the surface.   Identification of water mites is
challenging (Hopkins, 1961; Soar & Williamson, 1925,
1927, 1929).

–   Body divided into two parts, the cephalothorax
    (consisting of the fused head and thorax) and the
    abdomen, the two parts joined by a slender waist
                                    spiders (order Araneae)

The two genera of spiders described here are foundnd
on or near the water, and can scuttle over the surface.
These are wolf spiders with legs ending in three
toothed claws (IV.2), and with eight eyes arranged in
three rows of unequal size (IV.3). Wolf spiders include
terrestrial and aquatic species which hunt their prey
on foot instead of spinning webs. Those that do not
correspond with the descriptions given below may be
land spiders which are accidentally on the water.
Interesting accounts of aquatic spiders are given in
Bristowe (1958).

Dolomedes belongs to the family Pisauridae, the
'nursery web' spiders. *Dolomedes fimbriatus* (Clerck)

(pl. 2.6) is a magnificent spider, perhaps the prize 'find' of the neuston. It 'could not be mistaken for another British species' (Locket & Millidge, 1951). Body length: females 13–20 mm, males 9–13 mm. Females are deep brown with two broad pale yellow or white bands along each side. There are small white spots on the abdomen in some specimens, usually males. The spherical egg-cocoon is carried by the mouthparts. This species is found in swamps, pools or streams which do not dry up; it climbs down plant stems and hides below the surface if disturbed. It is called the 'raft spider' in older accounts, but Bristowe (1958) prefers 'swamp spider'. Several may be found together. The species is widespread but local in the Cambridgeshire Fens and the 'mosses' of Cumbria and Lancashire, and locally fairly common in the southern counties of Britain.

The family Lycosidae includes the genus *Pirata*, marked on the cephalothorax with a 'V' pointing backwards. The four British species are found in damp, marshy places and hunt agilely on the water surface. *Pirata piraticus* (Clerck) is illustrated as the commonest species (pl. 2.5). The body is chocolate brown; the cephalothorax has a dark central line forking in front; the abdomen has two yellowish lines along it, and behind, two lines of white spots converging towards the rear. Body length 4.5–9 mm. It spins a white silken tubular shelter (IV.4) and stands at the mouth of the tube with its forelegs resting on the water surface to detect vibrations. Found commonly throughout Britain.

P. hygrophilus Thorell is also very common all over the country. *P. latitans* (Blackwall) and *P. piscatorius* (Clerck) are widespread but rare.

The lycosid spiders are covered with a thick pile of fine hairs, giving them a strongly water-repellent surface which enables them to run across the water. Lycosids are frequently found on the surface having dropped from overhanging vegetation, but many species of the genus *Pardosa* are associated with water, being found on marshy ground or on the margins of lakes and rivers.

For further identification of spiders see Jones (1983), which gives coloured photographs of many species, or Locket & Millidge (1951) and Locket, Millidge & Merrett (1974).

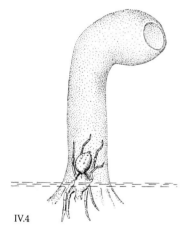

IV.4

V Key to Protozoa

After calm, still weather, a rich variety of microscopic plants and animals are sometimes found at the surface (p. 18). Very little is known about the microneuston, although many beautiful creatures are found in it. Some of them will be members of the plankton which have been captured at the surface. Most species found at the surface can be found in other habitats as well. We need to know which animals are the constant residents of the water surface.

This book does not offer a key to the genera of microneuston and cannot be used for critical identification of this group. Instead, it describes some examples of genera often found at the surface. Identification of rotifers and protozoans to species requires the skills and experience of a specialist. It also requires living specimens, so it is almost impossible to provide a specialist with voucher specimens. It is therefore often necessary to be satisfied with an incomplete identification, accompanied by descriptive notes and accurate drawings from life, with a scale line. Identification requires special techniques for handling and observation which are given in chapter 7.

The Protista is an enormous and varied group and its classification has undergone many transformations in recent years. The current taxonomic framework is given in 'An illustrated guide to the Protozoa' (Lee, Hutner & Bovee, 1985). For general purposes, Jahn, Bovee & Jahn (1979) and Kudo (1966) are still useful, and Curds (1982) and Curds, Gates & Roberts (1983) key out and describe all known genera of freshwater ciliated protozoans, with illustrations.

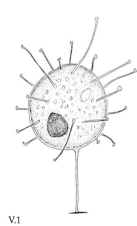

V.1

1   With numerous cilia (tiny hairlike structures which beat synchronously) on at least part of the body   6
–   Cells without cilia, naked or with a few whip-like flagella   2

2   With knobbed tentacles (V.1); average size 50–100 μm
                 subclass Suctoria (phylum Ciliophora)
    Example
    Sphaerophrya (pl. 7.2) is an unstalked form commonly found floating at the surface, where it preys on other small protists which are captured by the evenly-distributed tentacles. Other unstalked genera include *Holophrya* and *Solenophrya*. The lack of a stalk may be related to their floating habit. Forms with short stalks, such as *Podophrya* (pl. 7.3) and *Acineta*, are also found attached to floating debris.
–   Without knobbed tentacles
                       phylum Sarcomastigophora 3

3   Each cell with one or two whiplike flagella which propel the body by a rowing or spiral motion
                flagellates (subphylum Mastigophora) 4
–   Without flagella; moving by protoplasmic flow, either of whole body or by distinct pseudopodia; (cell naked or with an external test or shell)
                amoebae (subphylum Sarcodina) 5

4   Cells solitary
     Examples

> *Chilomonas* (pl. 4.7). Colourless; two unequal flagella attached inside a 'canal'; body contains an ovoid, flattened, colourless plastid and usually many polygonal starch grains. 30–50 μm.

> *Trachelomonas* (pl. 4.5). Body enclosed in a spherical or ovoid envelope which is often ornamented and which becomes dark brown with age and may make the plastids and eyespot difficult to see; locomotory flagellum emerges from a neck or collar. Widespread, particularly in peaty pools in reducing conditions. 20–45 μm.

> *Euglena* (pl. 4.1). Spindle-shaped when swimming; undergoing very characteristic changes of shape when stationary; with numerous bright green disc- or band-shaped plastids; one large flagellum is visible, emerging from a pocket (the reservoir) at the front; size varies with species, 20–500 μm long. *Euglena* may occur in enormous numbers forming a green crust on the surface of ponds after still dry weather. It rounds up to form a cyst or resting stage very readily (even while being examined on a slide) and large numbers of such cysts are often found on the water surface.

> *Phacus* (pl. 4.6). Flattened, asymmetrical body of constant shape, with prominent longitudinal or oblique ridges; large flagellum and red eye-spot; disc-shaped bright green plastids. 30–150 μm long.

–   colonial forms
     Examples

plastid: a cell inclusion, which may be coloured (often green, orange or yellow) or colourless

> *Synura* (pl. 4.4). Colony of 2–50 individuals; cells ovoid or pear-shaped, each with one or two golden-brown plastids and two short flagella, of more or less equal length, with which the colony swims. Colony 100–400 μm in diameter; cells 20–50 μm long.

> *Anthophysa* (pl. 4.3). Colony of many colourless individuals, each of which is very small (about 5–6 μm long) and has two unequal flagella. Each cluster of cells is on the end of a branching, twisted stalk which is colourless at first but becomes yellow-brown with age; a whole colony may detach from the stalk and swim about; detached individuals become amoeboid. Found in similar conditions to *Trachelomonas*.

5   Sarcodina
     Body housed in a test (shell or case)
     testate amoebae (class Lobosea, subclass Testacealobosia)

> Some testate amoebae can be identified in Corbet (1973) and Ogden & Hartley (1980).

Example

Arcella (pl. 4.2). Body contained in a beret-shaped test with a circular opening beneath, from which emerge a few blunt, finely granular pseudopodia; tiny gas bubbles may help to float the animal at the water surface; the test is yellow-brown at first, and darkens with age. *Arcella* is ubiquitous amongst sphagnum and other water plants and in bottom deposits, and its habitat range extends to the surface; it is one of the genera most constantly found there. Test 30–260 μm in diameter.

–   Body naked, with lobed or finely branched pseudopodia which extend slowly from the rounded or elongate body, which has a granular appearance

naked amoebae (class Lobosea, subclass Gymnamoebia)

A range of small (5–50 μm) amoebae, such as the genera *Mayorella* and *Hartmanella*, are very common in the hyponeuston, to which their small size and gliding locomotion make them well adapted. They are difficult to identify without prolonged study (see Bovee in Lee, Hutner & Bovee, 1985 or Page, 1976).

6   Ciliates (other than Suctoria)

Attached, often by a stalk, and with the cilia confined to a prominent ring winding anticlockwise towards the mouth opening

peritrichs (class Oligohymenophora, subclass Peritrichia)

Example

Vorticella (pl. 7.7). Bell-shaped body attached to an unbranched stalk which contains a contractile thread and which coils like a tight spring when contracted. Solitary. Vorticellids are often very abundant, attached to filamentous algae or to floating decayed leaf skeletons. Body length usually about 50–100 μm.

–   Free-swimming or creeping form, not attached to substrate   7

7   Contained in a transparent case which may be covered with debris

tintinnids (order Choreotrichida, suborder Tintinnina)

Example

Tintinnidium (pl. 8.6). A conical or trumpet-shaped protozoan with a conspicuous ring of ciliary structures (membranelles) winding in a clockwise direction to the mouth. Body 40–200 μm. These delicate and elegant ciliates are, strictly speaking, planktonic, swimming while carrying the transparent case, inside which they are attached, but they are sometimes found at the surface.

–   Not enclosed in a case   8

8   Ciliates which seem to 'walk' on short 'legs' which
     consist of fused bundles of cilia (cirri)   9
–   Gliding forms without leg-like cirri   other ciliates
     Examples
     Paramecium (V.2). Slipper-shaped, with a long, broad
     mouth groove running obliquely from the front end
     back to the mouth in the hind half of the cell. Body
     uniformly covered with short cilia; 80–300 μm.
     Chilodonella (pl. 7.6). Body cilia restricted to ventral
     surface; mouth basket of about 12 rods.

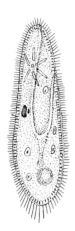

V.2

9   Body flattened; sides of body fringed with cirri
                              spirotrichs (class Spirotrichea)
     Examples
     Oxytricha (pl. 7.5). Body flexible from side to side;
     fringe of cirri is more or less continuous around rear
     edge. Length usually about 100–200 μm.
     Stylonychia (pl. 7.1). Body stiff; cirri not fringing rear
     edge, but there are three large cirri at the rear end.
     Length usually about 100–300 μm.
–   Body domed, with ridges along the back; sides of body
     not fringed with cirri
                              hypotrichs (subclass Hypotrichia)
     Examples
     Euplotes (pl. 7.4). Well developed membranelles
     associated with 'mouth' and extending halfway or
     more towards rear of body. Length usually at least
     80 μm.
     Aspidisca (pl. 7.8). Membranelles associated with
     'mouth' inconspicuous; no cirri fringing margins or at
     rear end. Length usually less than 50 μm.
     The spirotrichs and hypotrichs are amongst the most
     widespread and successful ciliates, ubiquitous in
     distribution but always abundant where there is some
     enrichment of the water by organic materials, either
     decaying products of the habitat or pollution. Their
     method of locomotion enables these protozoans to use
     the underside of the surface film much as they would
     use any other surface, so their presence in the neuston
     is probably fortuitous.

Guide to some rotifers (phylum Rotifera) found at the surface

Rotifers are microscopic animals. A few reach a
maximum length of 2 mm, but most are much smaller.
Size within a species can vary greatly with differing
environmental conditions, but males are generally
smaller than females. Males may be as little as 40 μm in
length, and they have reduced internal structure. They
are rarely found and are not covered by the key and
descriptions given here. Many common planktonic

genera appear in surface collections, especially if there is a good supply of food such as unicellular algae (less than 20 μm), bacteria and detritus. Some examples of genera found regularly in surface collections are described below, but these notes cannot be used for critical identification. Pontin (1978) and Ruttner-Kolisko (1974) give keys and illustrations for planktonic rotifers only; see also Voigt (1978).

It is often necessary to watch the animals for some time before the foot and crown of cilia are extended.

Without a foot
Examples

Filinia (pl. 6.5). Body sack-shaped, with three or four very long thin movable spines, usually two at the front and one at the back; two red eyes. Length up to 250 μm.

lorica: a transparent, sometimes flexible shell of thickened cuticle which maintains body shape and may bear spines, ridges, facets or other ornamentation.

Keratella (pl. 6.10). Body without movable spines or processes; body wall stiff, forming a distinct lorica consisting of a dorsal plate with distinct facets and with up to six spines on the front margin; ventral plate finely granulated (rough). Length up to 350 μm.

With a foot but no toes
Example

Collotheca (pl. 6.6). Foot long, slender, and very contractile; animal deformable, without a lorica, living in a wide transparent gelatinous sheath; crown of cilia forming a big, thin-walled funnel surrounding the mouth. Length 500 μm or more.

With a foot bearing toes at the end
Examples

Rotaria (pl. 6.8). Cuticle very flexible, allowing leech-like extension and retraction of body and foot; extended foot very long and with three toes and two spurs; crown of cilia retracted during movement. Length can exceed 1500 μm when extended.

Pontin (1978) records *R. neptunia* (Ehrenberg) as a sporadic migrant in the plankton of shallow muddy waters.

Brachionus (pl. 6.4). Foot long and very flexible, retractable, ringed; toes small or minute; body flattened dorsoventrally; lorica usually having two, four or six well developed spines on the front margin and sometimes spines at the rear as well. A very variable genus with many species. Lorica 200–600 μm long.

Squatinella (pl. 6.9). Two toes; lorica thin, transparent, rounded on the back, with three spines at the rear;

crown of cilia covered by a semicircular transparent shield which is not retractile. Length up to about 220 μm.

Euchlanys (pl. 6.7). Lorica domed, pear-shaped or egg-shaped, deeply notched behind; ventral plate flat; dorsal and ventral plates connected by a deep fold; foot short, not retractable, with two or three sections; toes sword-shaped; only one eye. Length up to about 340 μm.

7 Techniques and approaches to original work

7.1 Collecting and preservation

Neuston occurs on any free water surface. A rich source is found on still waters, amongst reeds and other vegetation on ponds, ditches and the edges of lakes, and quiet backwaters of streams and rivers. Some epineuston, such as whirligig beetles, larger pondskaters and some flies frequent the open water of lakes, and of streams where there is an appreciable current.

Microneuston will not accumulate if there is a current or wind, creating surface waves, although wind and current may aggregate surface material in sheltered places. The best method for collecting microneuston is to use a stout glass sheet about 30 cm square, which can be pushed in at right angles to the surface film and withdrawn slowly, so that the film adheres to both sides. This is then scraped off into a shallow dish using a neoprene windscreen scraper. The film can be concentrated by moving the glass through the water perpendicular to the surface and withdrawing it at an angle, underneath the film. Where this method is not practicable, a large-bulbed wide pipette is useful to suck up surface material, or a 'pooter' (suction bottle, fig. 27) with two long rubber tubes, one of which goes to the collector's mouth, while the other, ending in a wide nozzle, is held parallel to the water surface. All water samples are best tipped out as soon as possible into chemically clean, shallow dishes such as Petri dishes, and left for an hour or two for the organisms to orientate themselves again in relation to the surface.

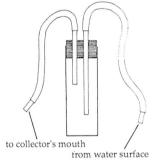

to collector's mouth

from water surface

Fig. 27. A 'pooter', used to collect surface film.

Small blocks of polyurethane sponge (plastic foam) may be used to collect protozoa from open water. They can be tied to a line strung across the water and held at the surface by floats, and left for a few days. Samples are collected by squeezing them over a jar (Cairns & Ruthven, 1980). Cellulose filter discs can be used in the same way. Very small samples of a few drops may be adequate. It is important to keep small samples separately to retain the community structure of tiny micro-communities. Slides or coverslips may be embedded horizontally into large corks, leaving a free surface to float on the water for varying periods of time. They may then be examined directly under the microscope. Petri dishes (glass, not plastic) may be fitted into polystyrene floats and placed upside down on the water surface. They are also easy to examine. These methods alter the habitat by providing a solid surface in place of the air–water interface so that atypical attached forms may be collected. Results should be checked against collections taken from the open surface.

The arthropods are collected at different levels. Flies which cruise just above the water are caught in a net of strong cloth, swept back and forth just above the water. A long-handled, D-shaped pond net is needed for large spiders, pondskaters, whirligig beetles and water boatmen, captured on or just below the surface. For small crustacea a fine-meshed bolting-silk hand-net of 180 meshes/in (7.2 meshes/cm) will catch even small stages of waterfleas, as well as dipteran larvae and pupae and the larger rotifers. Individual small animals such as spiders, *Velia* and *Microvelia* can be collected with a pooter or scooped up on a white plastic spoon (white so that small animals are more easily seen). Small forms like *Hebrus*, common amongst sphagnum moss, are found most easily by putting a handful of moss, together with some water, into a small washing up bowl and shaking out the moss in the water. The animals are more easily seen here than in the field.

Disney and others (1982) found that white water traps were very effective in catching Diptera, particularly the neustonic *Hydrellia modesta*. The traps consisted of a white bowl, 31 cm in diameter, half filled with water to which several drops of detergent were added. Other colours of bowl were tested but the largest numbers were caught in white ones. The presence of water was thought to be more significant than the type of bowl, but it must be borne in mind that a white bowl is a prominent and artificial feature in the habitat which may trap abnormally large numbers of flies compared with their normal distribution on a natural water body.

To investigate dispersal of winged pondskaters, Landin & Vepsäläinen (1977) used light-reflecting glass traps. As well as gerrids, large numbers of beetles such as *Helophorus* were caught, and no doubt other insects attracted to a light-reflecting surface will also be trapped. The trap consisted of an oblong sheet of glass, about 60 x 100 cm, held in a wooden frame with sides 2 cm high. The glass sloped at an angle of about 6° to the ground. The lower and shorter edge of the glass was placed over a collecting trough, containing water and a little detergent. Insects landing on the glass usually slid down into the trough. Traps may be placed near water, and emptied at suitable intervals. If they are to be left for several days, weak formalin may be added to the water to preserve any captive insects.

Because of the way they jump, springtails are easily trapped in a Petri dish coated inside with vaseline and floated on a suitable habitat. Small insects like *Hebrus* and springtails may be captured by hand in a dry tube. Rare or protected creatures such as *Dolomedes* should be left undisturbed, or if necessary, collected carefully for examination and then released again alive in the same location from where they were collected.

Many epineustonic animals such as spiders, pondskaters and beetles are shy, and well-equipped with sensory receptors so that they quickly detect naturalists as well as predators and take evasive action! It is necessary to stalk them with a minimum of disturbance; or having disturbed them, to wait patiently for their return to their preferred stretch of water.

All samples should be labelled immediately with the date and place of collection, with labels written in pencil or waterproof ink and placed inside the tubes. To preserve wet specimens for examination, they should be dropped immediately into 70% alcohol, contained in 3 in by 1 in glass specimen tubes. Otherwise they may be kept alive for further examination or culture.

For examination and storage, bugs, beetles and the larger flies may be killed by exposure to ethyl acetate vapour in a corked glass specimen tube and then mounted on insect pins. Most large insects are pinned through the middle of the thorax. Beetles are normally pinned through the right wingcase, or mounted with water-soluble glue (such as gum tragacanth) onto neat slips of white card. Small flies are pinned obliquely (say, from the top right to the bottom left of the thorax) with tiny headless pins (A1 size) onto a strip of polyporus or polyethylene foam, the shape and size of half a matchstick. The insect (or the card or foam strip on which it is mounted) should be mounted fairly high on a number 12 insect pin; below it on the same pin goes a label giving the date and place of capture and other relevant information. Pinned insects can be stored in a box with a close-fitting lid to exclude mites. Cork-lined wooden insect boxes are good but expensive. A clear plastic sandwich box with a tight-fitting floor of polyethylene foam is a satisfactory alternative.

Very small or delicate flies lose legs and antennae easily if pinned, and are better mounted on microscope slides with a mountant such as Berlese's fluid or polyvinyl lactophenol. Again, labelling is essential.

7.2 Sampling

A compound microscope is essential for the study of the microneuston. It may be possible to arrange to borrow one from a sympathetic school, at least during school holidays.

For rough quantitative samples of Protista and rotifers, a pipette delivering a standard-sized drop is used to place a drop of surface water on a slide. The drop is covered with a coverslip and examined using a compound microscope with a mechanical stage. Organisms occurring in small numbers are counted by scanning the whole area of the coverslip. Another useful method is to float several coverslips placed randomly on the surface of a sample for a standard length of time (say, one hour) and then put the coverslips onto clean slides, and count the relevant species, again using the mechanical stage to scan the whole coverslip. For denser populations, the slide is viewed through a squared eyepiece and a standard number of squares sampled.

The problems of sampling invertebrates adequately are discussed by Disney and others (1982). For a complete survey, it is best to use a variety of different sampling techniques, used at different times throughout the year and, if possible, for several years. These authors point out that a selective approach is more valuable than spreading one's efforts more thinly.

It is difficult to devise methods for absolute quantitative sampling. Larger species may be counted directly. Quadrats may be devised to suit the situation, such as metre square quadrats made in the form of a wooden frame floated on the water, or tripod-mounted binoculars defining a field of view. The numbers of particular species are recorded for each quadrat sample. A random sampling system (see Lewis & Taylor, 1967) must be used so that the averaged results of several quadrat counts give a true picture of the distribution of a species within that particular habitat. Southwood (1978) discusses 'nearest neighbour' techniques for estimating population densities of surface insects.

7.3 Handling microneuston

However collected, the material should be placed in shallow dishes to form a thin layer of water with a large surface area and allowed to settle for an hour or two, so that the film can reform on the surface. Square thin coverslips (preferably no. 1) are then carefully floated on the surface. After a further period of an hour or more they are removed with forceps, and dropped onto a slide and examined immediately. Drops may also be removed with a fine teat pipette. Light-sensitive forms such as flagellate protozoa and rotifers will collect on the lighter side of a dish lit from one side; this is a useful method of concentrating these forms.

It is necessary to identify both protozoa and rotifers from live material. Preservation distorts them and makes them unrecognisable. They may be slowed down using methyl cellulose solution (Methocell or Polycell wall-paper paste). 10 g is soaked in 45 ml boiling water for about 30 min, and then mixed with 45 ml cold water to produce a clear, viscous solution. A drop of water to be examined is placed in a ring of solution on a slide and a coverslip added. The animals are slowed down as they swim into the viscous solution. Another method is to place a few strands of cotton wool on the slide before adding the water drop, and then a coverslip. To anaesthetise ciliate protozoa, 0.01% nickel sulphate or 1% copper sulphate solutions may be used, although different species respond differently to these and to methyl cellulose solutions. Much of the internal structure of micro-organisms can be seen by cutting down the illumination or using phase contrast microscopy, if available.

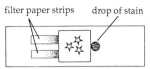

filter paper strips   drop of stain

Fig. 28. Staining live
specimens beneath a
coverslip.

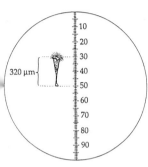

320 μm

Fig. 29. Using a micrometer
eyepiece.

Fig. 30. *Stentor*, a large
ciliate protist.

Stains or preservatives should be floated under the coverslip by placing a drop of solution at one side, and drawing it through slowly using small strips of filter paper applied to the other side (fig. 28). Protists may be stained with 0.5% methyl green in 1% acetic acid, to stain the nuclei. Lugol's iodine (4 g iodine, 6 g potassium iodide in 100 ml deionised water) will show up cilia and flagella. Jahn, Bovee & Jahn (1979) is a very useful guide to techniques for Protozoa.

Having a very thin layer of water under the coverslip will serve to slow down many rotifers. They may be preserved while fully extended using 1% formalin (toxic!) or 35% alcohol. More drastic preservatives cause them to retract the foot or crown of cilia and to buckle the lorica (case). Preserved rotifers are used to check details of lorica shape, spines or jaws. Donner (1966) gives good practical instructions for studying rotifers.

Drawings of live specimens are the only way of making permanent records of protozoa and rotifers. Simple line drawings are required with no shading, recording as accurately as possible the relative size and shape of all recognisable features. Drawings may be made on record cards and built up into a card index of animals recorded. As far as possible, accurate measurements must be added to any records.

Objects seen under the microscope are measured in micrometres (μm). A micrometre is 1/1000 of a millimetre. In the absence of special micrometers, rough estimates of size can be made from the size of a micro-organism relative to the diameter of the whole field of view. For example, if a student microscope has a 10x eyepiece and a 10x low power objective, the total magnification is 100x and the field diameter is about 1600 μm. A rotifer, say, which had a length of one quarter of a diameter of the field of view must be about 400 μm long. (A useful standard for comparison is a red blood cell which always has a diameter of 8 μm.) For accurate measurements, an eyepiece containing a micrometer scale is used. The scale may divide the diameter of the field into 100 parts. Each small division of the micrometer is then 16 μm. The scale is lined up beside the animal and its length measured (fig. 29). For higher magnifications, the relative size of the divisions must be calculated in relation to the magnification of the lenses used (see Jahn, Bovee & Jahn, 1979) or the micrometer can be calibrated against a micrometer slide, which bears a 1 mm scale with 100 divisions. Variability of size within a species allows an accuracy of about 10% to be adequate for identification of Protista and Rotifera. Sizes are recorded by adding a scale-line to each drawing, with the length in μm. Neustonic Protista are usually less than 1000 μm long, but a few ciliates are 1–3 mm long, just visible to the naked eye. Of these, *Stentor* is sometimes found at the surface (fig. 30). Female rotifers may be anything between 100 and 1000 μm in length; males are much smaller, often only about 1/4 of the length of females. Rotifer colonies may be larger, up to about 4 mm long.

Micro-organisms may be kept for several hours on a slide by sealing off the edges of the coverslip with petroleum jelly (vaseline) to prevent drying up. A thin smear is applied on the slide in a square, roughly the size of the coverslip, and a drop of water placed inside the square. The coverslip is carefully lowered over the vaseline square. Such slides may be kept for several days if placed in a damp chamber (fig. 31). Cavity slides, if available, are ideal to make minicultures of this kind. Mixed cultures of protozoa, rotifers, bacteria and algae may be kept for weeks at a time in covered shallow dishes if the water is topped up from time to time, preferably with clean rain water. The composition of these isolated communities can change very rapidly. Forms which were not present originally hatch out of cysts, resting eggs and suchlike, so only those species known to be present originally in an active state can be regarded as characteristic of the habitat from which the sample came. Nevertheless, such cultures are very useful for making more detailed records of forms which have been few in number, but which multiply in the dishes. Observations on feeding and life cycles may be made.

Samples may be enriched by the addition of culture media which encourage growth of food organisms. Algal growth is promoted by the addition of a few drops of a solution of inorganic salts such as Knop's* solution; bacterial growth is aided by the presence of one or two boiled wheat or rice grains. Weak infusions of organic substances have long been used to culture micro-organisms. 1–2 g hay or dried lettuce in a litre of deionised water, boiled and filtered, or 5 g garden soil, boiled, allowed to settle and the solution filtered, give solutions suitable for levels of bacterial growth which will not kill the larger organisms by lack of oxygen or production of excessive waste products.

As protozoa and rotifers feed on bacteria, algae, each other and members of their own kind, mixed cultures are a valuable source of material. Rapid growth of bacteria is followed by population outbursts of ciliates feeding on them, followed in turn by predatory ciliates and rotifers. Rotifers are often the last to appear and may survive for weeks after the growth of most other types has declined. Soil extract promotes slow growth of diatoms and desmids, as well as more rapid bacterial growth.

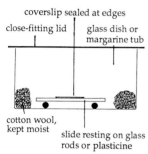

coverslip sealed at edges

close-fitting lid   glass dish or margarine tub

cotton wool, kept moist   slide resting on glass rods or plasticine

Fig. 31. Keeping slides of live material in a damp chamber.

*Knop's solution (modified, after Pringsheim)
   g per litre of deionised water

KNO_3	1.0
$Ca(NO_3)_2$	0.1
K_2HPO_4	0.2
$MgSO_4.7H_2O$	0.1
$FeCl_3$	0.001

7.4 Keeping neuston

Many of the neustonic arthropods may be maintained in cultures for further study using very simple methods. Aquatic bugs (*Velia, Mesovelia, Hydrometra* and *Gerris*) may be kept in small containers, perhaps with a little piece of pondweed on which to lay their eggs (although these may also be laid on the walls of the dish). It is preferable, if possible, to use pond or rain water, rather than tap water. The water needs changing regularly to prevent the accumulation of surface scum. Once eggs are laid, it is better to remove the adults. *Velia* and *Gerris* in particular are cannibalistic. To study nymphs, they must be separated from the adults. Ideally, each insect needs a separate container, or large aquaria containing growing plants which provide some cover may be used. Small bugs like *Hebrus, Microvelia* and *Velia* may be kept in plastic sandwich boxes containing damp vegetation which must be kept moist. Sphagnum is ideal.

Notonecta can also be kept in aquaria containing water plants, but they need aeration of the water, and a muslin or wire gauze cover to prevent them from flying away. Winged forms of *Gerris* will not fly if the temperature is kept low (below about 12 °C).

It is easy to provide food for all these predators. Fruit flies (*Drosophila*) are an excellent food source, readily available in most school laboratories; the vestigial-winged forms are best as live prey for obvious reasons. If cultures are not available, wild *Drosophila* are easily captured by placing jam-jars containing a little mashed banana in the boughs of trees. The flies are attracted to the fruit and cultures may be started by bringing in the jars and covering them with cotton lids. Mashed, ripe banana and a pinch of dried yeast is an adequate food for new cultures. The flies breed rapidly; their larvae burrow into the food and the life cycle is completed in about 2 weeks. They may be killed by placing a glass tube containing some flies over a hot radiator, or gently heating the tube with a match. Freshly killed flies are fed to the predators a few at a time. Houseflies, mosquito larvae and waterfleas may also be provided.

Mosquito and midge larvae may be reared in shallow enamel dishes containing water enriched with nutrient sources such as diluted milk or milk powder, hay infusion or wheat infusion, with a little dried yeast added. Old algal or protozoan cultures may also be fed to them. All of these media promote growth of micro-organisms and are also suitable for culturing waterfleas.

One of the problems with using infusions of various kinds is that growth of micro-organisms is uncontrolled and may lead to deleterious changes which kill off the animals feeding on the micro-organisms. A useful, simple apparatus which might be modified for studying the neuston food chain is the 'Nuffield Biology' experimental food chain (fig. 32, somewhat modified). Three tanks are set up (30 x 30 x 60 cm is a suitable size). The first tank can contain hay infusion, or an infusion of horse manure. The manure should be 7–10 days old, but not dry or mouldy. A layer filling about 1/3 of the tank is covered with tap water and allowed to stand for several days, after which it is inoculated with pond water and left for another 2–3 days. The other two tanks must be filled to the same level as the first with tap water. The second is inoculated with waterfleas. The three tanks are connected by tubes acting as siphons. It is desirable to have two tubes connecting a pair of tanks in case one gets blocked. When setting up the apparatus, the connecting tubes are filled with water, making sure that all air is excluded. Bubbles from the aerator pump cause a slow flow of water through the three tanks so that the water is continuously circulated.

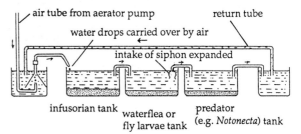

Fig. 32. An experimental food chain (after fig. 127 in Nuffield Biology Text IV, *Living Things in Action* (1966), Longman Group UK Ltd, by kind permission of the Nuffield–Chelsea Curriculum Trust).

The advantages of the apparatus are that small quantities of food travel from tanks 1 and 2, preventing the development of foul surface conditions (to which forms like *Gerris* are sensitive) in tanks 2 and 3, and the animals in tank 2 are not exposed to the full effects of decomposition in tank 1. It may be modified in various ways. Tank 3 could be omitted and epineustonic predators introduced on the surface of tank 2. Having three tanks may make it possible to devise ways of measuring the food supplied to the predator tank. Varying the rate of bubbling will regulate not only the rate of water flow, but also the food supply from one tank to the next. Tank 2 might be inoculated with fly larvae, remembering that the tank then needs covering with muslin to prevent any hatching adults from escaping. The same precaution is necessary if investigating winged predators such as *Notonecta* or *Gerris*. Greasing the walls of the tank prevents wingless bugs from escaping.

The springtail *Podura* will feed on the duckweed *Lemna*, kept in small aquaria with 5–8 cm of water (Noble-Nesbitt, 1963b). Individual specimens could be kept in separate tubes if required (fig. 33).

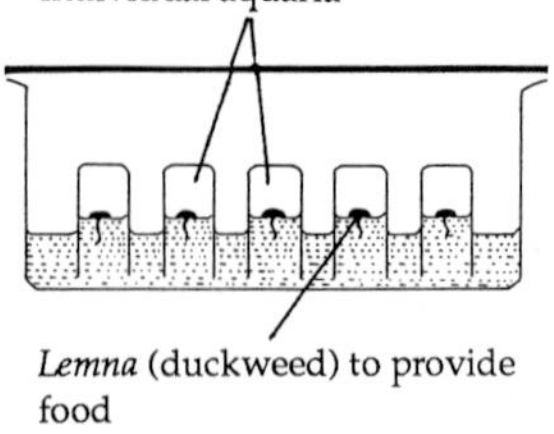

inverted specimen tubes make individual aquaria

Lemna (duckweed) to provide food

Fig. 33. Rearing *Podura* in small aquaria (after Noble-Nesbitt, 1963b).

7.5 Exploring surface effects

Standard methods of measuring surface tension require special apparatus and most values given in published tables apply only to pure liquids. Even exposure to the atmosphere deposits a film of lipid, and natural bodies of water contain large amounts of dissolved and suspended materials. Comparisons of the relative surface properties of different bodies of water can be made by investigating surface pressure – the difference between the surface tension of the water surface and that of pure water (72.7 mN m^{-1} at 20 °C). This can be done very simply by testing the spreading power ('spreading pressure') using drops of mixtures of fatty substances (Adam, 1937). Small amounts of surfactants containing hydrophilic groups are mixed with a pure hydrocarbon oil to give a graded series of mixtures of different spreading pressures.

   Adam used n-dodecyl alcohol mixed with a water-white non-spreading mineral oil. A suitable oil available today is Shell lubricating oil, Dialar oil BG. Adam gives the following values of surface pressures against which the drops would spread slowly on seawater, to the nearest mN m^{-1}.

% dodecyl alcohol	Surface pressure against which slow spreading occurs (mN m^{-1})
0.05	Very slow spreading, even on clean water
0.07	1
0.1	3
0.2	6
0.3	12
0.5	16
0.7	19
1.0	22

Beament (unpublished information) used solutions of terpineol in BDH liquid paraffin (specific gravity 0.84) for the same purpose. The surface pressures against which drops of these solutions will just spread are given below.

ml terpineol/100 ml solution	surface pressure (mN m^{-1} at 20 °C)
0.1	4.5
0.2	7.5
0.5	13.0
1.0	18.0
2.0	21.0
5.0	25.0
10.0	30.0
terpineol	38.0

To spread on a surface, the applied surfactant must have a spreading pressure greater than that of the natural surface. Starting with the weaker solutions, a drop is put onto the water surface. If it remains as a lens-shaped drop, the natural film has a greater spreading pressure. The next concentration is tried. The surface pressure of the film on the natural surface is equal to that of the applied surfactant solution which will just spread.

   This simple method makes possible comparisons between various natural bodies of water, and could help to throw light on the extent to which the distribution of neuston is related to variations in surface tension, whether natural or caused by pollutants. The technique has been used to discover what concentration of an expensive surfactant should be incorporated into an oil film to control mosquitoes on infested swamps.

A simple test for pollution by detergents is the froth test. A water sample of about 50 ml is placed in a corked bottle of about 100 ml capacity, and shaken for half a minute. Measure how long it takes for the froth that forms to break and disappear. Clean water samples produce little or no froth, and large bubbles disappear within 3–5 seconds. Relatively large quantities of detergent cause increasing frothing, the breaking time of which can be anything from 10 seconds to 18 hours, and the disappearance of all bubbles takes from 4 hours to 18 hours or longer. Times taken are directly related to the amounts of detergent present.

A natural surface, seen from various angles, may show 'rainbow' colours. This may be due to a layer of mineral oil many molecules thick, but it may also be produced by dense concentrations of diatoms or flagellates like *Euglena*. If the water is in sunlight, look at the reflection through a piece of polaroid sunglasses while rotating it. The light reflected from a clean water surface is polarised and the picture will change as the polaroid is rotated. If it does not, there is probably a surface contaminant which may be many molecules thick.

7.6 Exploring surface properties of organisms

For experiments, tap water direct from the mains is very free of surfactant and is normally preferable to deionised water. Wash all apparatus used very thoroughly in running tap water, and be careful not to touch any part you use with your fingers. Surfactants spread on solid surfaces and at the air–water interface, so the important area of any container to have absolutely clean is the part in contact with the water surface.

A clean water surface can be obtained by filling a container under the tap and allowing it to overflow for a minute or so. Any surfactant will be spilt with the water. To be sure that instruments such as needles, forceps or pipettes are free of surfactant, dip them repeatedly *through* the water surface while it is spilling. Stubborn surfactants can usually be removed by gentle flaming.

To detect surfactants in water samples, sprinkle a little chemically clean talc ('French chalk') from a pepper pot over the clean water. Do not add so much that it forms a solid platform. Add a drop of water sample gently to the centre of the talc. (Do not drop it from a height, or it will make a hole mechanically.) A monolayer of surfactant will push the chalk back leaving a clear area. This technique allows detection of even minute amounts. The rapidity with which the area spreads gives some indication of spreading pressure. To check that instruments are free of contamination, dip them in and out through the talc-covered surface.

To investigate whether live organisms produce surfactants, a clean water surface lightly dusted with talc is prepared, as above. If, for example, an insect is then allowed to move over the surface, and secretes surfactant, it will leave a permanently clear zone in the talc. Insects which are normally submerged may be introduced beneath the surface from a clean pipette. If they surface and produce surfactant, the latter will push the talc back, leaving a clear area after the insect submerges again, showing that it is not merely due to mechanical disturbance of the talc. If 'holes' appear in the talc above a submerged insect, is it grooming and releasing surfactant droplets which rise to the surface?

Measuring contact angles (θ) can indicate whether a surface is hydrophilic or hydrophobic. Clean water is taken up from below the spilt surface of tap-water into a clean, fine pipette, preferably newly-made by drawing out a piece of glass tubing in a flame, taking suitable safety precautions. A water droplet of up to 1–2 mm diameter is placed on the surface to be investigated. The profile of the droplet must be viewed horizontally, to measure the contact angle. Some types of binocular dissecting microscopes may be pivoted horizontally on their stand. The best system is to use a prism for viewing. Using a matt black background and side illumination, one can, with practice, estimate the contact angle of droplets by eye to ± 5 degrees. Use cold light. Focussed strong light will heat the surface and melt normally solid lipids, interfering with the result.

Valuable information is obtained simply by determining whether particular areas are readily wetted (water spreads over the surface), just wetted ($\theta = 70–90°$), hardly wetted (θ just greater than 90°), or very hydrophobic (θ about 120°). Boundaries between areas with different surface properties can often be detected by moving a tiny drop around while it is still attached to the end of a pipette. The drop will bounce off unwettable areas but cling to wettable ones.

Always test whether an insect on which you make contact angle measurements has surfactant on its surface. If surfactant gets onto the droplets used, the measurements will be seriously affected. The surface properties of structures like a leg can be seen by removing the leg, taking care not to smear the surface with any of the internal tissues which contain very surface-active substances. The limb is lowered slowly through a clean water surface sprinkled with a little talc. Suitable illumination helps to show how the meniscus rises on hydrophilic areas and is depressed by hydrophobic regions. Continue observation while slowly lifting the limb out of the water.

If its surface properties are to be investigated, an insect has to be immobile, but it should not be anaesthetised with organic vapours like ether and ethyl acetate because these dissolve and may disorientate surface lipids. It is better, if possible, to immobilise the insect with carbon dioxide and decapitate it. Avoid at all costs contamination with its body fluids or secretions, or grease from your fingers. Surface properties are also totally altered by abrasion of the cuticle. Always use fresh material; surfaces can change after death, especially if they dry out.

7.7 Presentation of results

Writing up is an important part of a research project, particularly when the findings are to be communicated to other people. A really thorough, critical investigation that has established new information of general interest may be worth publishing if the animals on which it is based can be identified with certainty. The *School Science Review* welcomes papers on biological topics; the *Journal of Biological Education* publishes material with an educational slant; the *Entomologist's Monthly Magazine* and the *Bulletin of the Amateur Entomologists' Society* publish short papers on insect biology. Those unfamiliar with publishing conventions are advised to examine current numbers of these journals to see what sort of thing they publish, and then to write a paper along similar lines, keeping it short, but presenting enough information to establish the conclusions. It is then time to consult an appropriate expert who can give advice on whether and in what form the material might be published. It is an unbreakable convention of scientific publication that results are reported with scrupulous honesty. Hence it is essential to keep detailed and accurate records throughout the investigation, and to distinguish in the write-up between certainty and probability, and between deduction and speculation. In many cases, it will be necessary to apply appropriate statistical techniques to test the significance of the findings. A book such as Parker's (1979) *Introductory Statistics for Biology* or the *OU Project Guide* (Chalmers & Parker, 1986) will help, but this is an area where expert advice can contribute much to the planning, as well as the analysis, of the work.

Further reading

Finding books

Some of the books and journals listed here will be unavailable in local and school libraries. It is possible to make arrangements to see or to borrow such works in several ways: by seeking permission to use the library of a local university or of the Royal Entomological Society of London, or by asking your local public library to borrow the work (or a photocopy of it) for you via the British Library, Document Supply Centre. This may take several weeks and it is important to present your librarian with a reference that is correct in every detail. References are acceptable in the form given here, namely the author's name and date of publication, followed by (for a book) the title and publisher or (for a journal article) the title of the article, the journal title, the volume number, and the first and last pages of the article.

Asterisks mark publications available from The Richmond Publishing Co. Ltd., P.O. Box 963, Slough, SL2 3RS.

Handbooks for the Identification of British Insects are published by the Royal Entomological Society of London, and can be bought from the British Museum (Natural History), Cromwell Road, London, SW7 5BD.

Freshwater Biological Association Scientific Publications are available from the Association, Institute of Freshwater Ecology, The Windermere Laboratory, Ambleside, Cumbria, LA22 0LP.

References

Adam, N.K. (1937). A rapid method for determining the lowering of tension of exposed water surfaces, with some observations on the surface tension of the sea, and of inland water. *Proceedings of the Royal Society* B **122**, 134–139.

Adam, N.K. (1951). *Physics and Chemistry of Surfaces.* Oxford University Press.

Andrews, A.R. & Floodgate, G.D. (1974). Some observations on the interactions of marine protozoa and crude oil reserves. *Marine Biology* **25**, 7–12.

d'Assis Fonseca, E.C.M. (1968). Diptera Cyclorrhapha Calyptrata section (b) Muscidae. *Handbooks for the Identification of British Insects* 9 (4b), 1–119. Royal Entomological Society of London.

d'Assis Fonseca, E.C.M. (1978). Diptera Orthorrhapha Brachycera Dolichopodidae. *Handbooks for the Identification of British Insects* 9 (5), 1–90. Royal Entomological Society of London.

Aveyard, R. & Haydon, D.A. (1973). *An Introduction to the Principles of Surface Chemistry.* Cambridge: Cambridge University Press.

Baier, R.E. (1972). Organic films on natural waters; their retrieval, identification and modes of elimination. *Journal of Geophysical Research* **77**, 5062–5075.

Balfour-Browne, F. (1950). *British Water Beetles.* Vol.2. London: Ray Society.

Barger, W.R. & Garrett, W.D. (1970). Surface active organic material in the marine atmosphere. *Journal of Geophysical Research* **75**, 4561–4566.

*Barnes, R.S.K. (ed.) (1984). *A Synoptic Classification of Living Organisms.* Oxford: Blackwell Scientific Publications.

Baudoin, R. (1955). La physico-chimie des surfaces dans la vie des Arthropodes aeriens, des miroires d'eau, des rivages marins et lacustres et de la zone intercotidale. *Bulletin Biologique de la France et de la Belgique* LXXXIX, 16–164.

Beament, J.W.L. (1976). The ecology of cuticle. In *The Insect Integument* (ed. Hepburn, H.R.). Elsevier Scientific Publishing Company.

Beament, J. & Corbet, S.A. (1981). Surface properties of *Culex pipiens pipiens* eggs and the behaviour of the female during egg-raft assembly. *Physiological Entomology* **6**, 135–148.

*Belcher, H. & Swale, E.M.F. (1976). *A Beginner's Guide to Freshwater Algae.* London: HMSO.

Benfield, E.F. (1972). A defensive secretion of *Dineutes discolor* (Coleoptera: Gyrinidae). *Annals of the Entomological Society of America* **65**, 1324–1327.

Berry, W.O. & Brammer, J.D. (1977). Toxicity of water-soluble gasoline fractions to fourth-instar larvae of the mosquito *Aedes aegypti* (L.). *Environmental Pollution* **13**, 229–234.

Berry, W.O., Brammer, J.D. & Bee, D.E. (1978). Uptake of water-soluble gasoline fractions and their effects on oxygen consumption in aquatic stages of the mosquito *Aedes aegypti* (L.). *Environmental Pollution* **15**, 1–22.

Bleckmann, H. & Rovner, J.S. (1984). Sensory ecology of a semi-aquatic spider (*Dolomedes triton*). *Behavioral Ecology and Sociobiology* **14**, 297–312.

Brindle, A. (1960). The larvae and pupae of the British Tipulinae (Diptera, Tipulidae). *Transactions of the Society for British Entomology* **14**, 63–114.

Brindle, A. (1967). The larvae and pupae of the British Cylindrotominae and Limoniinae (Diptera, Tipulidae). *Transactions of the Society for British Entomology* **17**, 151–216.

Brinkhurst, R.O. (1959a). Alary polymorphism in the Gerroidea (Hemiptera–Heteroptera). *Journal of Animal Ecology* **28**, 211–230.

Brinkhurst, R.O. (1959b). A description of the nymphs of British *Gerris* species (Hemiptera–Heteroptera). *Proceedings of the Royal Entomological Society of London* **A 34**, 130–134.

Brinkhurst, R.O. (1960). Studies on the functional morphology of *Gerris naias* Degeer (Hemiptera–Heteroptera Gerroidea). *Proceedings of the Zoological Society of London* **133**, 531–559.

Bristowe, W.S. (1958). *The World of Spiders.* New Naturalist **38**. London: Collins.

Brocher, F. (1910). Les phénomènes capillaires. Leur importance dans la biologie aquatique. *Annales de Biologie lacustre* **4**, 89–138.

Bronmark, C., Malmqvist, B. & Otto, C. (1984). Anti predator adaptations in a neustonic insect (*Velia caprai*). *Oecologia (Berlin)* **61**, 189–191.

Cairns, J. Jr. & Ruthven, J.A. (1980). Artificial microhabitat size and the number of colonising protozoan species. *Transactions of the American Microscopical Society* **89**, 100–109.

*Chalmers, N. & Parker, P. (1989). *OU Project Guide: Field Work and Statistics for Ecological Projects.* London: Open University/Field Studies Council.

*Chinery, M. (1976). *A Field Guide to the Insects of Britain and Northern Europe.* London: Collins.

*Chinery, M. (1986). *Collins Guide to the Insects of Britain and Western Europe.* London: Collins.

Chvála, M. (1975). The Tachydromiinae (Dipt. Empididae) of Fennoscandia and Denmark. *Fauna Entomologica Scandinavica* **3**, 336 pp.

Chvála, M. (1983). The Empidoidea (Diptera) of Fennoscandia and Denmark II. *Fauna Entomologica Scandinavica* **12**, 281 pp.

Cockrell, B.J. (1984). Effects of temperature and oxygenation on predator–prey overlap and prey choice of *Notonecta glauca*. *Journal of Animal Ecology* **53**, 519–532.

Collin, J.E. (1958). A short synopsis of the British Scatophagidae (Diptera). *Transactions of the Society for British Entomology* **13**, 37–56.

Collin, J.E. (1961). *British Flies. Empididae.* Cambridge: Cambridge University Press.

Colyer, C.N. & Hammond, C.O. (1968). *Flies of the British Isles* (2nd edn). London: Frederick Warne.

*Corbet, S.A. (1973). An illustrated introduction to the testate rhizopods in Sphagnum, with special reference to the area around Malham Tarn,Yorkshire. *Field Studies* **3 (5)**, 801–838.

Cranston, P.S .(1982). *A Key to the Larvae of the British Orthocladiinae (Chironomidae)*. Freshwater Biological Association Scientific Publications, **45**.

Cranston, P.S., Ramsdale, C.D., Snow, K.R. & White, G.B. (1987). *Key to the Adults, Male Hypopygia, Fourth-instar Larvae and Pupae of the British Mosquitoes (Culicidae) with Notes on their Ecology and Medical Importance.* Freshwater Biological Association Scientific Publications, **48**.

*Croft, P.S. (1986). A key to the major groups of British freshwater invertebrate animals. *Field Studies* **6(3)**, 531–579.

Crow, S.A., Ahearn, D.G., Cook, W.L. & Bourquin, A.W. (1975). Densities of bacteria and fungi in coastal surface films as determined by a membrane-absorption procedure. *Limnology and Oceanography* **20**, 644–646.

*Curds, C.R. (1982). *British and other Freshwater Ciliated Protozoa. Part I.* Synopses of the British Fauna **22**. Linnean Society.

*Curds, C.R., Gates, M.A. & Roberts, D.M. (1983). *British and other Freshwater Ciliated Protozoa. Part II.* Synopses of the British Fauna **23**. Linnean Society.

Darnhofer-Demar, B. (1969). Sur Fortbewegung des Wasserlaüfers *Gerris lacustris* L. auf der Wasseroberflache. *Zoologischer Anzeiger Supplement* **32**, 429–439.

Disney, R.H.L. (1975). *A Key to British Dixidae.* Freshwater Biological Association Scientific Publications, **31**.

*Disney, R.H.L., Erzinçlioğlu, Y.Z., Henshaw, D.J. de C., Howse, D., Unwin, D.M., Withers, P. & Woods, A. (1982). Collecting methods and the adequacy of attempted faunal surveys, with reference to the Diptera. *Field Studies* **5**, 607–621.

Donner, J. (1966). *Rotifers* (trans. H.G.S. Wright). London: Frederick Warne.

Dumont, H.J. & Pensaert, J. (1983). A revision of the Scapholeberinae (Crustacea: Cladocera) *Hydrobiologia* **100**, 3–45.

Erlandsson, A., Malmqvist, B., Andersson, K.G., Hermann, J. & Sjörström, P. (1988). Field observations on the activities of a group-living semi-aquatic bug, *Velia caprai. Archiv für Hydrobiologie* **112 (3)**, 411–419.

Evans, G. (1975). *The Life of Beetles.* London: George Allen & Unwin.

Fairbairn, D.J. (1986). Does alary polymorphism imply dispersal dimorphism in the water-strider, *Gerris remigi? Ecological Entomology* **11**, 355–368.

*Fitter, R. & Manuel, R. (1986). *Collins Field Guide to Freshwater Life.* London: Collins.

Fjellberg, A. (1980). *Identification Keys to Norwegian Collembola.* Oslo: Norwegian Entomological Society.

Foelix, R.F. (1982). *Biology of Spiders.* Cambridge, Mass., and London: Harvard University Press.

Fowler, W.W. (1887). *The Coleoptera of the British Islands.* Vol. 1. London: Reeve & Co.

Freeman, P. (1983). Sciarid flies (Diptera, Sciaridae). *Handbooks for the Identification of British Insects* **9 (6)**, 1–68. Royal Entomological Society of London.

*Friday, L.E. (1988). A key to the adults of British water beetles. *Field Studies* **7(1)**, 1–151 (An AIDGAP key).

Fryer, G. (1956). A cladoceran, *Dadaya macrops* (Daday) and an ostracod *Oncocypris mülleri* (Daday) associated with the surface film of water. *Annals and Magazine of Natural History* **Series 12, IX**, 733–736.

Gilby, A.R. (1980). Transpiration, temperature and lipids in insect cuticle. In *Advances in Insect Physiology* (ed. Berridge, M.J., Treherne, J.E. & Wigglesworth, V.B.), Vol. 15, 1–34.

Giller, P.S. & McNeill, S. (1981). Predation strategies, resource partitioning and habitat selection in *Notonecta* (Hemiptera/Heteroptera). *Journal of Animal Ecology* **50**, 789–808.

Goldacre, R.J. (1949). Surface films on natural bodies of water. *Journal of Animal Ecology* **18**, 36–39.

Green, J. (1963). Seasonal polymorphism in *Scapholeberis mucronata* (O.F.M.). *Journal of Animal Ecology* **32**, 425–439.

Guthrie, D.M. (1959). Polymorphism in the surface water bugs (Hemiptera–Heteroptera, Gerroidea). *Journal of Animal Ecology* **28**, 141–152.

Hackman, R.H. (1974). Chemistry of the insect cuticle. In *Physiology of Insecta*, Vol. VI (ed. Rockstein, M.), pp. 216–270. New York and London: Academic Press.

Hadley, N.F. (1981). Cuticular lipids of terrestrial plants and arthropods; a comparison of their structure, composition and waterproofing function. *Biological Reviews* **56**, 23–47.

Harde, K.W. (1984). *A Field Guide in Colour to Beetles*. London: Octopus Books.

Harding, J.P. & Smith, W.A. (1974). *A Key to the British Freshwater Cyclopid and Calanoid Copepods*. Freshwater Biological Association Scientific Publications **18**.

Holdgate, M.W. (1955). The wetting of insect cuticles by water. *Journal of Experimental Biology* **32**, 591–617.

*Hopkins, C.L. (1961). A key to the water mites (Hydracarina) of the Flatford area. *Field Studies* **1(3)**, 45–64.

Hyman, L.H. (1951). *The Invertebrates*. Vol. 3. Acanthocephala, Aschelminthes and Entoprocta. New York: McGraw Hill.

*Jahn, T.L., Bovee, E.C. & Jahn, F.F. (1979). *How to know the Protozoa* (2nd edn). Iowa: Brown & Co.

Jahn, W. (1972). Ecological investigations of ponds with special regard to the consequences of water pollution by oil. *Archiv für Hydrobiologie* **70**, 442–483. (German with English summary.)

Johnson, B.D. (1976). Non-living organic particle formation from bubble dissolution. *Limnology and Oceanography* **21**, 444–446.

*Jones, D. (1983). *The Country Life Guide to Spiders of Britain and Northern Europe*. Feltham: Newnes.

Jones, J.G., Horne, J.E., Moorhouse, P. & Powell, D.L. (1980). *Petroleum Hydrocarbons in Freshwaters. A Preliminary Desk Study and Bibliography*. Freshwater Biological Association Occasional Publications **9**.

Kudo, R.R. (1966). *Protozoology* (5th edn). Springfield, Illinois: Thomas.

Landin, J. & Vepsäläinen, K. (1977). Spring dispersal flights of pond skaters: *Gerris* spp. (Heteroptera). *Oikos* **29**, 156–160.

Lang, H.H. (1980). Surface wave discrimination between prey and non-prey by the back-swimmer *Notonecta glauca*. *Behavioral Ecology and Sociobiology* **6**, 233–236.

Laurence, B.R. (1952). Observations on *Hydrellia (Hydropota) griseola* Fln (Diptera Ephydridae). *Entomologist's Monthly Magazine* **88**, 31–33.

Laurence, B.R. (1953). On the feeding habits of *Clinocera (Wiedemannia) bistigma* Curtis (Diptera:Empididae). *Proceedings of the Royal Entomological Society* **A 28**, 139–144.

Lee, J.L., Hutner, S.H. & Bovee, E.C. (ed.) (1985). *An Illustrated Guide to the Protozoa*. Kansas, U.S.A.: Society of Protozoology.

Lewis, T. & Taylor, L.R. (1967). *Introduction to Experimental Ecology*. London: Academic Press.

Linsenmaier, K.E. & Jander, R. (1963). 'Entspannungsschwimmen' von *Velia* und *Stenus*. *Naturwissenschaften* **50**, 231.

Locke, M. (1974). The structure and formation of the integument of insects. In *Physiology of Insecta*, Vol. VI (2nd edn) (ed. Rockstein, M.). New York and London: Academic Press.

Locket, G.M. & Millidge, A.F. (1951). *British Spiders* Vol. I. London: Ray Society.

Locket, G.M., Millidge, A.F. & Merrett, P. (1974). *British Spiders* Vol. III. London: Ray Society.

Macan, T.T. (1959). *A Guide to Freshwater Invertebrate Animals*. London: Longmans.

Macan, T.T. (1976). *A Key to British Water-bugs (Hemiptera–Heteroptera)*. Freshwater Biological Association Scientific Publications 16 (2nd edn).

McCauley, R.N. (1966). The biological effects of oil pollution in a river. *Limnology and Oceanography* 11, 475–486.

MacIntyre, F. (1974). The top millimetre of the ocean. *Scientific American* 230(5), 62–77.

McMullen, A.I., Reiter, P. & Phillips, M.C. (1977). Mode of action of insoluble monolayers on mosquito pupal respiration. *Nature, London* 267, 244–245.

Marshall, J.F. (1938). *The British Mosquitoes*. London: British Museum (Natural History).

Matthey, W. (1971). Ecologie des insectes aquatiques d'une tourbière du Haut-Jura. *Revue Suisse de Zoologie* 78 (2), 267–536.

Merritt, R.W. & Cummins, K.W. (1978). *An Introduction to the Aquatic Insects of North America*. Dubuque, Iowa: Kendall/Hunt.

Miall, L.C. (1934). *The Natural History of Aquatic Insects*. London: Macmillan.

Milne, L. J. & Milne, M. (1978). Insects of the water surface. *Scientific American* 238(4), 134–142.

Mitis, H. von (1937). Ökologie und Larventwicklung der Mitteleuropäischen *Gerris* Arten. *Zoologische Jahrbücher* 69, 337–372.

Murphey, R.K. & Mendenhall, B. (1973). Localisation of receptors controlling orientation to prey by the back-swimmer, *Notonecta undulata*. *Journal of Comparative Physiology* 84, 19–30.

Nachtigall, W. (1961). Funktionelle Morphologie, Kinematik und Hydromechanik des Ruderapparates von *Gyrinus*. *Zeitschrift für Vergleichende Physiologie* 45, 193–226.

Nachtigall, W. (1974). Locomotion: mechanisms and hydrodynamics of swimming in aquatic insects. In *Physiology of Insecta*, Vol III, (2nd edn) (ed. Rockstein, M.), pp. 381–432. New York and London: Academic Press.

Noble-Nesbitt, J. (1963a). Transpiration in *Podura aquatica* L. (Collembola, Isotomidae) and the wetting properties of its cuticle. *Journal of Experimental Biology* 40, 681–700.

Noble-Nesbitt, J. (1963b). The cuticle and associated structures of *Podura aquatica* at the moult. *Quarterly Journal of Microscopical Science* 104, 369–391.

Nummelin, M. & Vepsäläinen, K. (1988). Does size and abundance distribution of food explain habitat segregation among developmental stages of water striders (Heteroptera: Gerridae)? *Annales Entomologici Fennici* 54, 69–71.

Ogden, C.G. & Hartley, R.H. (1980). *An Atlas of Freshwater Testate Amoebae*. Oxford: Oxford University Press.

Page, F.C. (1976). *An Illustrated Guide to Freshwater and Soil Amoebae*. Freshwater Biological Association Scientific Publications 34.

*Parker, R.E. (1979). *Introductory Statistics for Biology* (Studies in Biology). London: Edward Arnold.

*Pinder, L.C.V. (1978). *A Key to Adult Males of British Chironomidae* (2 volumes). Freshwater Biological Association Scientific Publications 37.

Pitkin, B.R. (1988). Lesser dung flies (Diptera: Sphaeroceridae). *Handbooks for the Identification of British Insects* 10 (5c), 1–175. Royal Entomological Society of London.

Pontin, R.M. (1978). *A Key to British Freshwater Planktonic Rotifera*. Freshwater Biological Association Scientific Publications 38.

Portier, P. (1911). Recherches physiologiques sur les insectes aquatiques. *Archives de Zoologie* 8 (5), 80–379.

*Redfern, M. (1975). Revised field key to the invertebrate fauna of stony hill streams. *Field Studies* 4, 105–115.

Reynolds, C.S. & Walsby, A.E. (1975). Water blooms. *Biological Reviews* 50, 437–481.
Richards, O.W. & Davies, R.G. (1977). *Imms' General Textbook of Entomology*. London: Chapman and Hall.
Rozkošný, R. (1973). The Stratiomyioidea (Diptera) of Fennoscandia and Denmark. *Fauna Entomologica Scandinavica* 1.
Ruttner-Kolisko, A. (1974). Planktonic rotifers: biology and taxonomy. *Die Binnengëwasser* XXVI (1).
Savage, A.A. (1989). *Adults of the British Aquatic Hemiptera Heteroptera: a Key with Ecological Notes*. Freshwater Biological Association Scientific Publications 50.
Savory, T. (1964). *Arachnida*. London and New York: Academic Press.
Scott, M.A. & Murdoch, W.W. (1983). Selective predation by the back-swimmer *Notonecta*. *Limnology and Oceanography* 28, 352–366.
Scourfield, D.J. (1894). Entomostraca and the surface film of water. *Journal of the Linnean Society, Zoology* 25, 1–19.
Scourfield, D.J. (1900). Note on *Scapholeberis mucronata* and the surface film of water. *Journal of the Quekett Microscopical Club* Ser. 2: 7, 309–312.
Scourfield, D.J. & Harding, J.P. (1966). *A Key to the British Species of Freshwater Cladocera*. Freshwater Biological Association Scientific Publications 5.
Séguy, E. (1934). *Faune de France* 28. Muscidae Acalypterae et Scatophagidae. Paris: Lechevalier.
Sieburth, J.M. (1976). Bacteria in marine ecosystems. *Annual Review of Ecology and Systematics* 7, 259–285.
Sieburth, J.N., Willis, P.J., Johnson, K.M., Burney, C.M., Laudie, D.M., Hinga, K.R., Canon, D.A., French, F.N., Johnson, P.W. & Davis, P.G. (1976). Dissolved organic matter and heterotrophic microneuston in the surface microlayers of the North Atlantic. *Science* 194, 1415–1418.
Smith, K.G. & Empson, D. (1955). Note on the courtship and predaceous behaviour of *Poecilobothrus nobilitatus* L. (Dipt., Dolichopodidae). *British Journal of Animal Behaviour* 3, 32–34.
Smith, K.G.V. (in press). Diptera larvae. *Handbooks for the Identification of British Insects*. Royal Entomological Society of London.
Soar, C.D. & Williamson, W. (1925, 1927, 1929). *British Hydracarina*. Vols. 1–3. London: Ray Society.
Southwood, T.R.E. (1978). *Ecological Methods* (2nd edn). London: Methuen.
Southwood, T.R.E. & Leston, D. (1959). *Land and Water Bugs of the British Isles*. London: Warne.
Spence, J.R. (1986). Relative impacts of mortality factors in field populations of the waterstrider *Gerris buenoi* Kirkaldy (Heteroptera Gerridae). *Oecologia (Berlin)* 70, 68–76.
Streble, H. & Krauter, D. (1973). *Das Leben im Wassertropfen*. Stuttgart: Kosmos.
Sturdy, G. & Fischer, W.H. (1966). Surface tension of slick patches near Kelp beds. *Nature, London* 211, 951–952.
Thorpe, W.H. (1950). Plastron respiration in aquatic insects. *Biological Reviews* 25, 344–390.
*Unwin, D.M. (1981). A key to the families of British Diptera. *Field Studies* 5, 513–553 (An AIDGAP key).
*Unwin, D.M. (1984). A key to the families of British Coleoptera (and Strepsiptera). *Field Studies* 6, 149–197 (An AIDGAP key).
Viets, K. (1936). Wassermilben oder Hydracarina (Hydrachnellae und Halacaridae). *Tierwelt Deutschlands* 31 & 32, 574 pp.
Voigt, M. (1978). *Rotatoria: die Rädertiere Mitteleuropas*. Berlin, Stuttgart: Borntraeger. 673 pp.
Ward, A. (1985). *Experimenting with Surface Tension and Bubbles*. London: Dryad Press Ltd.

Wangersky, P.J. (1976). The surface film as a physical environment. *Annual Review of Ecology and Systematics* **7**, 161–176.

Wigglesworth, V.B. (1964). *The Life of Insects.* London: Wiedenfeld and Nicolson.

Wilcox, R.S. (1972). Communication by surface waves. Mating behaviour of a water strider (Gerridae). *Journal of Comparative Physiology* **80**, 255–266.

Some useful addresses

Suppliers of equipment

Entomological:
Watkins and Doncaster, Four Throws, Hawkhurst, Kent
Microscope and entomological:
GBI (Labs) Ltd, Shepley Industrial Estate, Audenshaw, Manchester, M34 5DW

Suppliers of biological books

New:
Richmond Publishing Co Ltd, P.O. Box 963, Slough, SL2 3RS
New and secondhand:
E.W. Classey Ltd, P.O. Box 93, Faringdon, Oxon, SN7 7DR

Societies

Amateur Entomologists' Society,
355 Hounslow Road, Hanworth, Feltham, Middlesex, TW13 5JH.
British Entomological and Natural History Society,
c/o The Alpine Club, 74a South Audley Street, London, W1Y 5FF
Royal Entomological Society of London,
41 Queens Gate, London, SW7 5HU
Quekett Microscopical Club,
c/o British Museum (Natural History), Cromwell Road, London, SW7 5BD
Freshwater Biological Association, Institute of Freshwater Ecology,
The Windermere Laboratory, Ambleside, Cumbria, LA22 0LP

Index